by RON & SUE FOURNIER

HPBooks

HPBooks
Published by the Penguin Group
Penguin Group (USA) Inc.
375 Hudson Street, New York, New York 10014, USA
Penguin Group (Canada), 90 Eglinton Avenue East, Suite 700, Toronto, Ontario M4P 2Y3, Canada
(a division of Pearson Penguin Canada Inc.)
Penguin Books Ltd., 80 Strand, London WC2R 0RL, England
Penguin Group Ireland, 25 St. Stephen's Green, Dublin 2, Ireland (a division of Penguin Books Ltd.)
Penguin Group (Australia), 250 Camberwell Road, Camberwell, Victoria 3124, Australia
(a division of Pearson Australia Group Pty. Ltd.)
Penguin Books India Pvt. Ltd., 11 Community Centre, Panchsheel Park, New Delhi—110 017, India
Penguin Group (NZ), 67 Apollo Drive, Mairangi Bay, Auckland 1311, New Zealand
(a division of Pearson New Zealand Ltd.)
Penguin Books (South Africa) (Pty.) Ltd., 24 Sturdee Avenue, Rosebank, Johannesburg 2196,
South Africa

Penguin Books Ltd., Registered Offices: 80 Strand, London WC2R 0RL, England

PRINTING HISTORY
HPBooks trade paperback edition / May 1989

Library of Congress Cataloging-in-Publication Data

Fournier, Ron.
Sheet Metal Handbook : how to form, roll, and shape sheet metal for competition, custom and restoration use / Ron and Sue Fournier.
p. cm.
Includes index.
ISBN 978-0-89586-757-5
1. Automobiles—Bodies—Design and construction. 2. Sheet-metal work. I. Fournier, Sue. II. Title.
TL255.F68 1989
629.2'6—dc19 88-30679
CIP

PRINTED IN THE UNITED STATES OF AMERICA

45 44 43 42

ABOUT THE AUTHORS

Ron Fournier's career as a metal fabricator spans over 25 years. He began with Holman and Moody in 1964, and has since that time worked for some of the greatest competitors in racing history. Teams such as Penske, A.J. Foyt, Kar Kraft, and Bob Sharp have all utilized Ron's unique skills to transform metal into various components for their championship-winning race cars. In the mid-Seventies, Ron founded Race Craft, which soon developed a nationwide reputation as one of the finest metal fabrication shops in the country. Today, he supervises the development of automotive prototype sheet metal projects for Entech Metal Fabrication in Troy, Michigan.

Ron's wife, Susan, is the reason Ron's years of fabrication experience can be translated into written form. After receiving a B.A. from Michigan State University, Sue went on to graduate work in English at West Chester State University in Pennsylvania and the University of Houston. Her publishing credits include a number of poems, short stories and a high school English textbook.

Ron and Sue's first joint effort, HP's *Metal Fabricator's Handbook,* received a Moto Award in 1984, the highest achievement possible in automotive journalism. Together, they create technical articles for the various car enthusiast publications across the country. It has been an unusual, yet productive joint effort. Over the years, Sue has learned a great deal about metal fabrication, while Ron's spelling has improved.

ACKNOWLEDGMENTS

Lujie Lesovsky deserves special thanks for two reasons. First, he taught me to appreciate metal fabrication at Holman and Moody's in 1964, which has much to do with why I've been perfecting this craft during the last 25 years. Second, he was kind enough to arrange visits to California metal shops so I could gather material for this book. Scott Knight, Mike Lewis, Marcel DeLey, Mel Swain, Fay Butler and Bill Bizer, all craftsmen in their own right, deserve thanks for taking the time to share with me their knowledge, which I am now passing on to you. Thanks to Gray Baskerville, of *Hot Rod Magazine* for being such a great West Coast connection.

Some businesses were especially helpful as well. Entech Metal Fabrication, my own home base, was extremely understanding whenever I took the time to document my own work and the work of others. Hydrocraft graciously allowed me to pose their large metal-working equipment for photographs, while T.N. Cowan supplied tools and materials for research.

Finally, I owe my two children, Nicole and Dan Fournier, many thanks for understanding that I had to forsake many hours of time with them to work on this book; and to my wife Sue, who is every bit as responsible for this book's completion as I am.

CONTENTS

Introduction................. 9

1 Getting Organized......... 11

2 Basic Hand Tools 15

3 Sheet Metal Equipment .. 27

4 Types of Sheet Metal..... 41

5 Patterns and Layout...... 49

6 Metal Shaping............. 55

7 Hammerforming........... 75

8 Riveting 89

9 Restoration 97

10 Sheet Metal Interiors.... 113

INTRODUCTION

Soon after the *Metal Fabricator's Handbook* was published in 1982, I was overwhelmed with letters requesting more specific information on working with sheet metal in particular. This book is in response to those demands. Though I thoroughly enjoyed writing the first book, I had much more fun this time. Why? Mainly because working with sheet metal gives me a great feeling of satisfaction. It is a highly creative and productive process, one that presents a new challenge to my skills with each individual project, even if it is one I've done before. Whether it's a dashboard, fender panel, transmission tunnel, wheel tub or hood scoop, I never tire of transforming a flat sheet of alloy into an object that is not only pretty to look at, but has some functional purpose as well. The end results are always worth the effort.

Like any highly specialized skill, working with sheet metal and learning to shape it into any form is a skill that requires many hours of practice, and patience. I can certainly give you the information you'll need to attain the skill, but I can't give you the desire. That can only come from you.

Before we can dive right in to creating a sheet metal project, there are some basic principles you need to know. We must first lay a foundation, one that we can build upon. The base of that foundation is the shop you are going to be working in. It must be properly organized for safety and to make working as easy as possible. It is also essential that you become thoroughly familiar

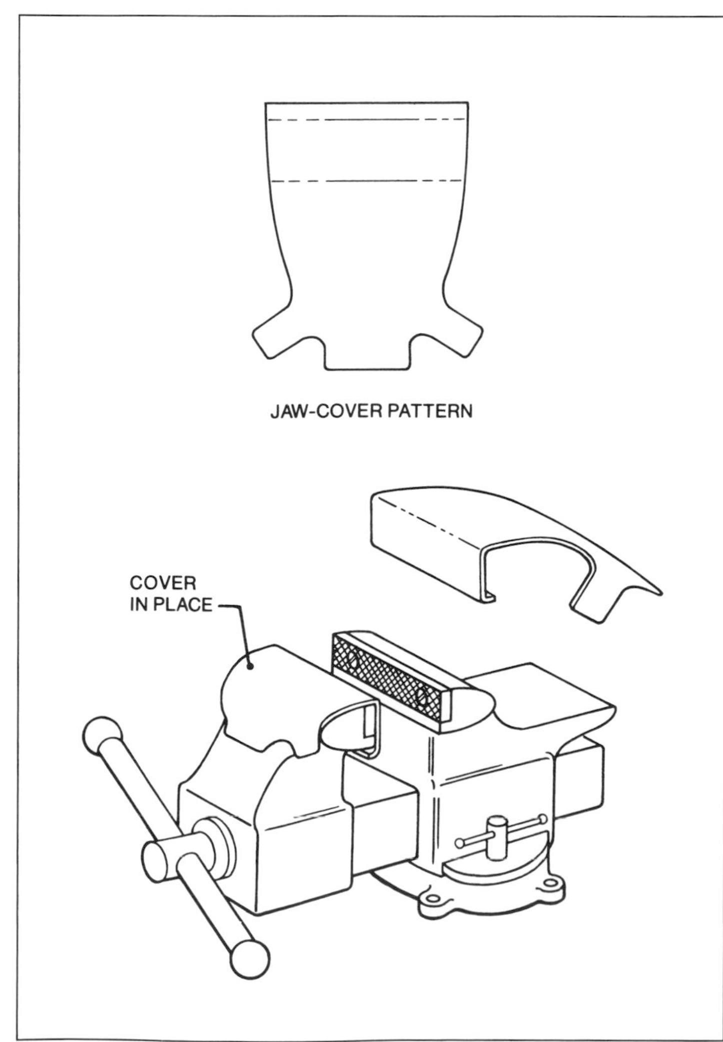

For immediate pay-off on a beginner project, vise jaw covers do the trick. They're easy to make and immediately useful, and keep from damaging your work in the vise. Use 3003-H14 aluminum, 0.063-inches thick, and the pattern as a guide to make your own.

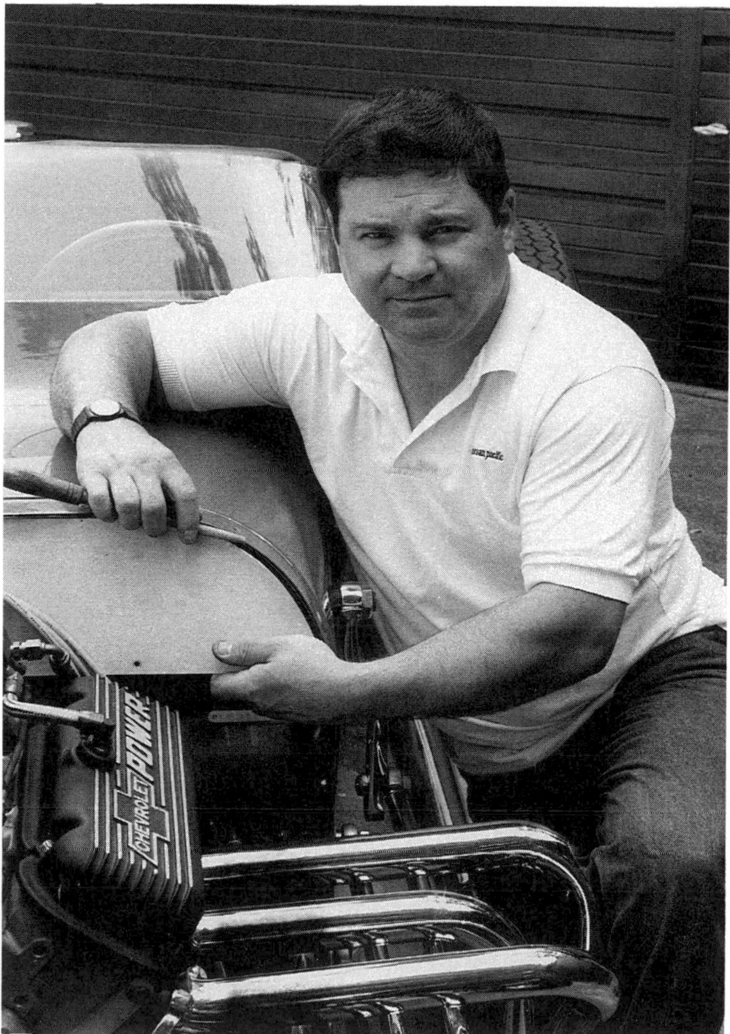

I've been working with sheet metal for over 25 years. Here, I'm applying the final touches to my roadster. All metal components were fabricated by me, using many of the methods that are outlined in this book.

with the most common hand tools, as well as the larger equipment used in sheet metal fabrication. Finally, you'll need to be briefed on the basic qualities of the different types of sheet metal available.

As I've already mentioned, working with metal requires a skill that can only be learned through *practice,* the next layer of our foundation. In the beginning, it is important to plan jobs at your own skill level. The most common mistake I've seen time and time again is someone taking on a project much larger than he can handle. This only leads to discouragement, and often failure. Some of the initial projects may seem too easy, but it's important to realize that each job is a steppingstone to larger, more rewarding projects as your skills improve and develop. The trick is to plan carefully what you will attempt. Is it a project you can handle? For instance, it would be a good idea to begin with a project requiring *layout* and *bending*, two of the most basic skills essential to metal work. Making a pair of vise jaw covers, for example, would be an excellent first project to test these skills. It requires minimum skill, practices layout, involves bending and results in a product that can be used again and again. A project doesn't have to be complex to be satisfying and useful.

A second phase project might be something requiring layout, bending and *riveting*. Making a dust pan doesn't sound all that thrilling, but it will polish skills for a more exciting project. In making a shop dust pan, you are using layout techniques, bending metal and learning about riveting. The product, the dust pan, is something you can use and take some pride in making.

A third level project would require layout, bending, riveting and *welding*. For instance, a hood scoop involves using all those skills. It is a project to be tackled only when your skill in each area is proficient.

A fourth level job would involve layout, bending, riveting, welding and perhaps several other different skills. The project might also include *metal shaping*, *metal finishing*, and using a *bead roller*. A project at this level is something you might attempt after successfully completing several jobs at the third level. Even then, it will take patience and perseverance to get good results.

By now, you are beginning to see the "snowball effect" of learning metal work. Each step along the way builds on the previous step, and each project builds on the base of the previous project. Master the basics, build up that foundation before you go on adding another skill with different tools or techniques. There are no short cuts.

Scott Knight, of Scott's Hammer Works in California, stands next to a Ferrari that he has reskinned using metal shaping techniques. This should give you some idea of the potential rewards to mastering metal shaping techniques.

CHAPTER

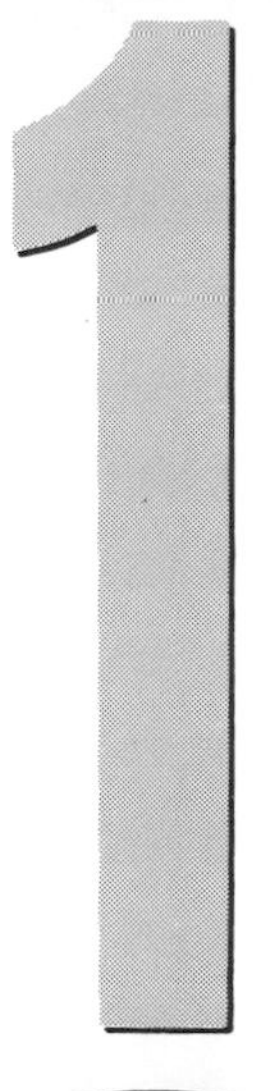

GETTING ORGANIZED

Organizing a sheet metal shop is crucial to success. It requires careful thought and planning. The shop doesn't necessarily have to be large, but it does need to be well organized. If it is well-planned, your work is more likely to progress smoothly and your chances for success will be increased.

An often overlooked area is proper *lighting*, which is important when working with sheet metal. Quite often, you are drawing fine lines then making cuts along those lines. The accuracy of those cuts often depends on how well you can see those lines. Natural light is ideal. Fluorescent light is okay if the fixtures are properly located. Lights need to be placed directly above the working areas, especially over the layout tables. The idea is to reduce or completely eliminate shadows. Shadows obscure some of the details of the work, and make it hard to be precise. So just don't throw up fixtures wherever it is convenient. Make sure that the shop you choose has the ability to be well-lit.

PLACING EQUIPMENT

Placement of equipment is another important factor. How easily you can use the equipment often depends upon how the equipment is arranged in the shop. Some equipment, like *beaders* and *rollers*, can be grouped together on a table or special work stand. It is much more convenient to keep these tools close to one another. Frequently, when you are using one of these tools, you need another one as well.

Some equipment should never be crowded together. The *shear* and *sheet metal brakes* need space around them. Metal shaping equipment, like *Kraftformers, English wheels, and power hammers*, also need lots of space. Use of shears, brakes and large shaping equipment often involves large panels of sheet metal. Obviously, you'll need plenty of working room to avoid any problems.

This is the shop I manage, Entech Metal Fabrication. Note how I've organized the equipment so there's plenty of working room. Also, check out the selective use of lighting. Photo by Michael Lutfy.

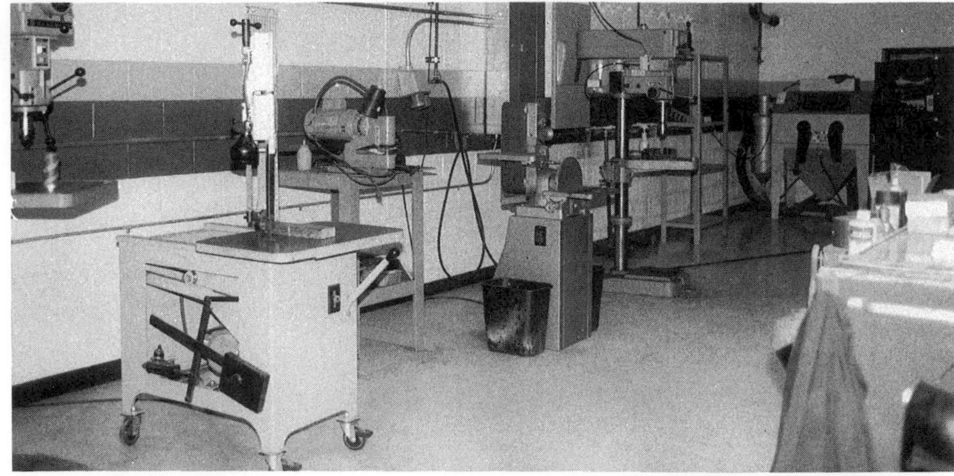

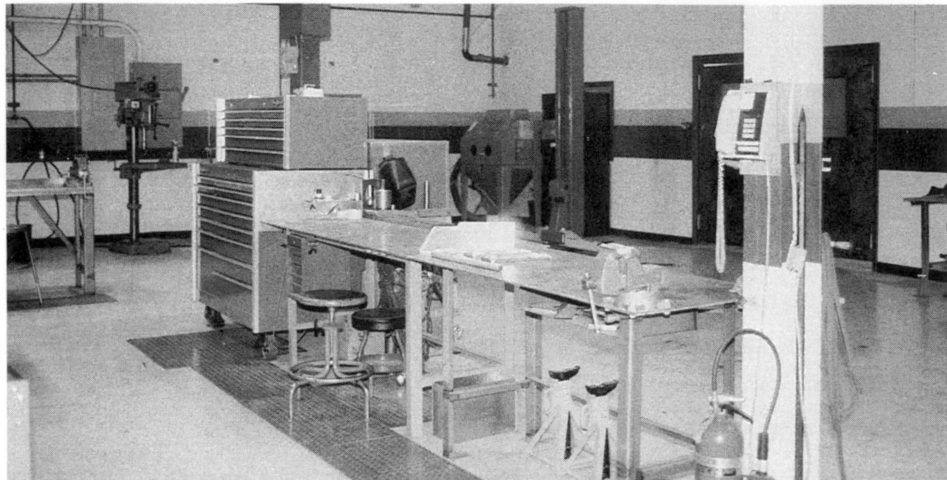

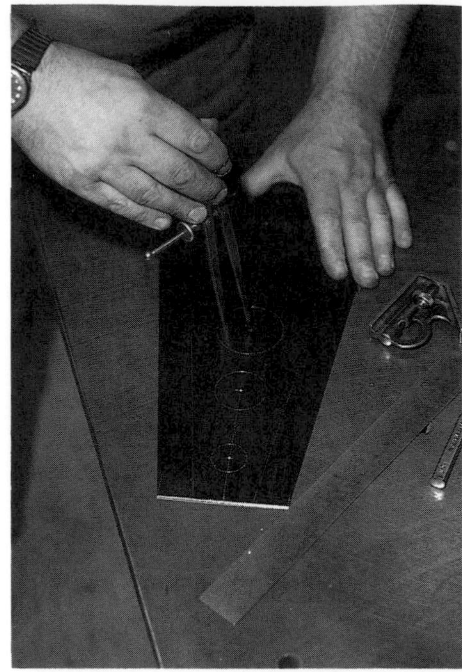

The layout bench is flat, and mine is made of heavy steel.

Allowing space around shop equipment has two great advantages: it lets you work conveniently and cleaning is simple. Areas where it is necessary to see lines clearly need to be well lit.

A *shear* needs plenty of room in front of it, where you will work. It also needs room behind it, because some metal pieces will extend through the shear and behind it while you are working. It is most convenient to have the shear placed with space around it in the first place. It is no small chore to rearrange the equipment after finding out your pieces are bumping into the rear wall. A *sheet metal brake* also needs space in front and back for the same reasons.

If you have enough shop space, it is best to allow plenty of room to walk between pieces of equipment. This spacious arrangement will make it more comfortable, and safer. It will also make it much easier to clean up the work area at the end of the day.

SHEET METAL BENCHES

Sheet metal benches can be categorized into two types. The first is the *layout bench*. It is usually large and made of either wood or metal, sometimes both. It is also very stable and heavy, and most importantly, it has a very flat and level top.

The top must also be very smooth and specifically made to prevent scratching the metal. Some top fabricators I know cover the top with a tightly woven, low-pile carpet to avoid scratching sheet metal as it is layed out. This is especially important if you're going to be working with aluminum frequently, which is soft and scratches easily. Another tip when setting up the layout bench is to put locking casters on the table so it can be easily moved and secured. This allows you to wheel the table to the work, rather than bring the work to the table. It also makes it easier to clean under the table.

The other common type of bench is the *work bench*. It differs from the layout bench in that it is smaller and much heavier, usually made of a very heavy metal. The working surface is usually no more than 38-inches above the floor. The height makes it easier and more comfortable to weld while sitting down.

The top is usually 3-1/2 X 6-feet square, and approximately 1/2-inch thick or more. This kind of table top offers enough support to perform any metal-working procedure: you can weld, grind, or hammer on it.

The work bench should have a large vise mounted on one corner. I wouldn't recommend a vise with jaws smaller than 4-inches, and it should always be *bolted* to the table, not clamped. I have seen people use C-clamps to hold a vise to a work bench. This creates a dangerous situation, because the vise could come loose.

A bench of this kind must also have very sturdy legs to withstand the pounding and grinding of heavy metal work. A lower shelf mounted about 10-inches from the ground and about 18-inches deep is very handy. The shelf is recessed enough to keep the table comfortable for seated welding,

This sheet metal rack is made from 0.060-inch steel tubing. It's strong enough to support a lot of weight and has rollers to make it mobile. Photo by Michael Lutfy.

Our steel tubing rack is made from 1/8-inch steel tubing for extra strength. It has rollers and is kept at least 3 feet from the wall for easy access. Photo by Michael Lutfy.

and offers some storage space.

The work bench looks good with a coat of paint, but don't paint the top surface. The paint would just be burned or chipped off. It could cause a bad ground when you are trying to weld something. It is a good idea to frequently apply a thin coat of oil on the top surface to prevent rusting.

SHEET METAL RACKS

Metal sheets, rods or bars need to be stored in special units, because they are large, heavy and awkward items. A proper rack or storage unit will keep them in good condition and easily accessible. Most shops choose to make their own steel racks, designed to fit around the shop and the materials it uses the most. A particular shop might use three kinds of *tubing*, for example. It would then design and build a rack to hold those three kinds of tubing. Another shop might use a wide variety of tubing, sheet and *bar stock*. Its rack would have to be more complicated to hold a wider variety of material.

Making the rack itself is not hard if you're proficient with a tape measure and basic welding techniques. The key is in the initial design. *Design it so the largest and heaviest pieces of metal will be stored at the bottom*. The smallest and lightest items should be stored at the top. The idea is to make a rack which will be stable when it is loaded with metal sheets, rods and bars. The rack could become lethal if it were loaded top-heavy.

The rack supports must be placed so metal will not hang off the rack or sag between the supports. It not only looks messy to have the metal hanging outside the supports, it is also dangerous. It is easy to cut yourself on a metal edge walking by. If the rack holds all the metal without overlapping the supports or sagging between supports it will look good, preserve the metal shape and be safer. As the rack may have to support a tremendous amount of weight, it should be made from heavy-gage, welded steel.

As for where to put the rack itself, right in the middle of the shop is one good idea, because all of the metal on the rack is easily accessible from all sides. However, be careful to keep it out of "traffic patterns," so people can walk past it without difficulty. Up against a wall is fine, as long as you leave enough room behind it to reach metal stored at the back. One place you must *never* store a rack loaded with stock sheet metal is near a door that leads outside. Wind will blow in rain or snow through the doorway and corrode the sheet metal, causing rust freckles that will have to be painstakingly removed before you can begin working on the metal. This is both time-consuming and irritating work, so avoid it by placing your racks in the proper place.

SHOP SAFETY

One final word about your shop—make sure that it is properly equipped to either avoid, or if necessary, handle emergencies. Working with sheet metal may not exactly qualify as "hazardous," but it can be dangerous, particularly to a novice who may not see a potentially harmful situation before it happens. Always equip your shop with several fire extinguishers that are fully charged and easily within reach. A misdirected welding torch can cause a flash fire in an instant. Make sure that the shop area is also well-ventilated. There will be times when you'll be working with flammable adhesives and chemicals, which should be stored securely in steel storage cabinets until they are used. Trash containers should be kept away from welding areas, and labeled as either "flammable" or "non-flammable." Another sound idea that is both profitable and ecological is to set aside a separate container for scrap aluminum to be recycled.

CHAPTER

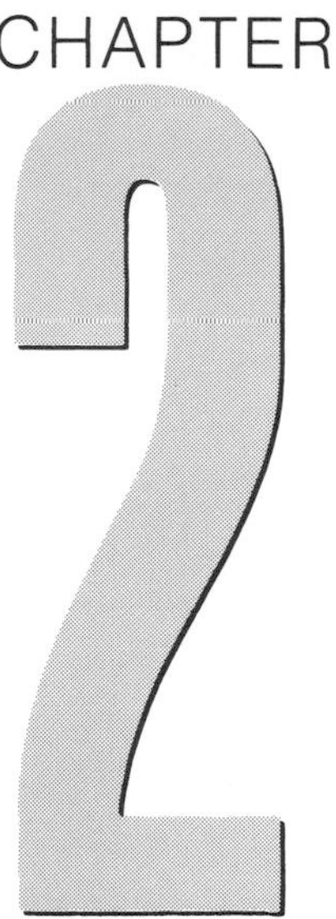

BASIC HAND TOOLS

Basic hand tools are fundamental to sheet metal work. All of the ones I've listed here will be used continuously. They are essential to performing even the most simple sheet metal project. I recommend that you not only purchase your own set of these tools, but that you also practice with them until you are proficient and fully comfortable using them. These are the tools that make it possible to measure, mark, cut and shape metal.

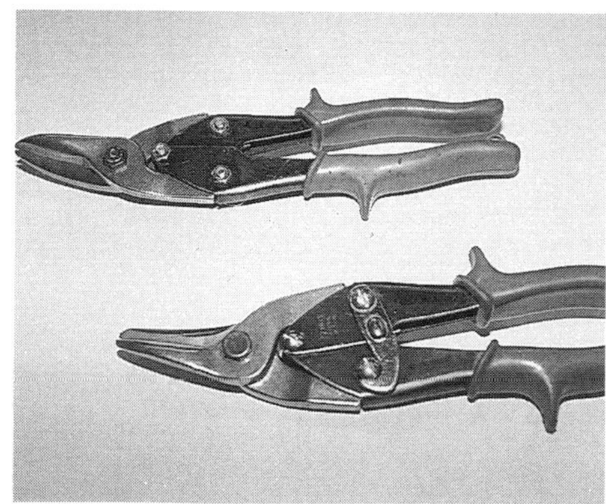

Every sheet metal worker needs a set of right and left cutting Aviation snips. They are the most common kind of snips used in sheet metal work. Use Tinner's snips for straight cutting, however.

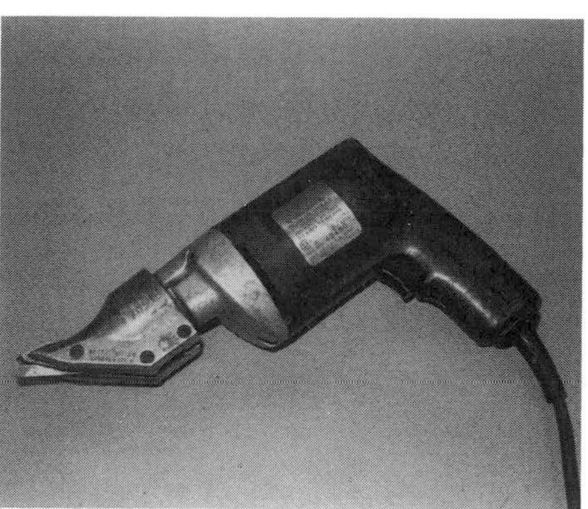

I couldn't begin to estimate how many hours and sore muscles these Kett electric metal shears have saved me. They make long, precise cuts—either curved or straight—and those cuts need only minimum clean-up.

A large low-speed electric drill is ideal for a sheet metal shop. I use this Sears drill with hole saws when I need to cut a hole and can't take the part to the drill press.

CUTTING TOOLS

These tools are simple, hand-held devices used to cut or trim small pieces of metal. They are essential items that should be in every sheet metal shop.

SNIPS

Snips are scissor-like devices with sharp blades used to cut metal. They are invaluable items that you'll use time and time again. There are basically two types of snips; *Aviation* and *Tinner's* snips. You'll need to purchase three pairs. One pair that cuts to the right, one pair that cuts to the left, and one pair that cuts a straight line. For right and left snips, I recommend the Aviation type because of their durability. For straight snips, however, I recommend the Tinner's straight snips over the Aviation straight snips. The Tinner's variety, which are normally used in the heating and air conditioning industry, make longer, nicer cuts. The straight Tinner's snips are

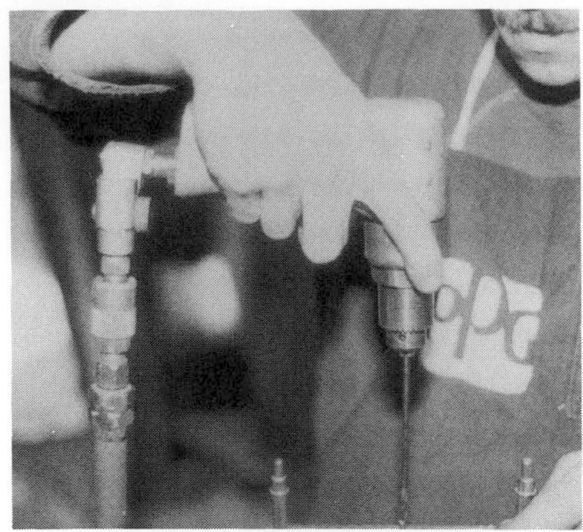

Pneumatic drills come in many sizes. This 1/4-inch capacity air drill is very compact. Some people prefer the 3/8-inch capacity air drill for the higher diameter of the chuck.

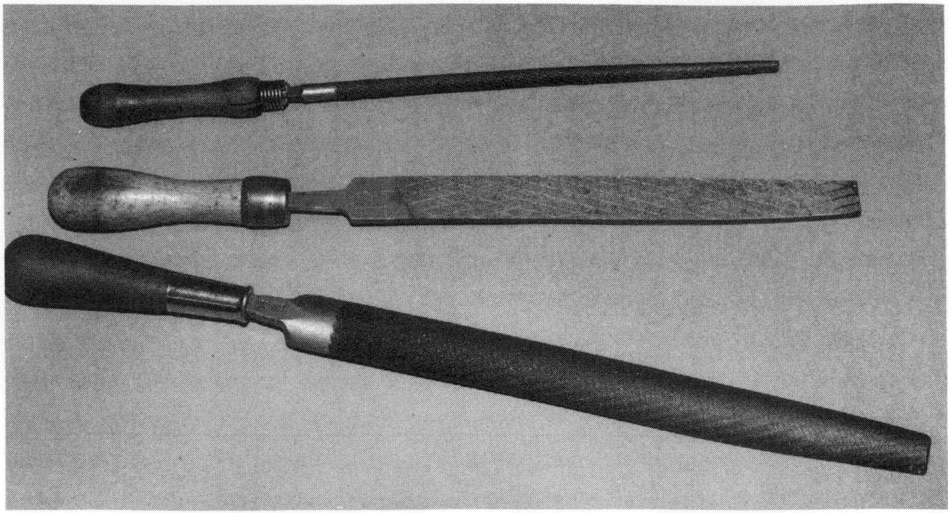

Sheet metal workers need a selection of files to smooth various surfaces and remove burrs from cut metal edges. Round, flat and half-round files are the most common.

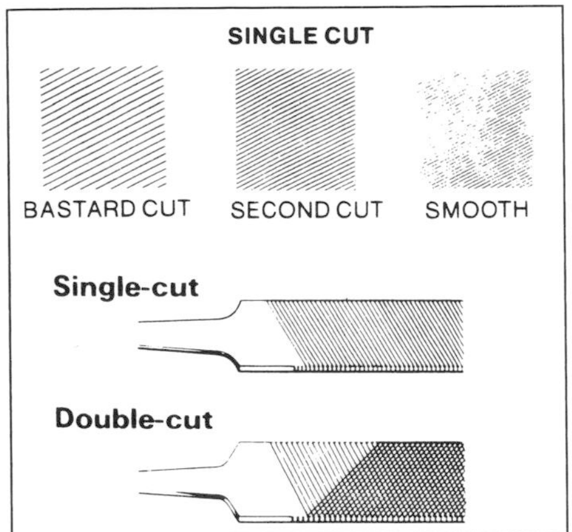

Most files are available with three grades of coarseness: bastard-cut, second-cut and smooth-cut in single- or double-cut. Bastard-cut file removes the most material with each pass; smooth-cut removes the least. Double-cut file is used with more pressure than the single-cut to remove material faster from work piece.

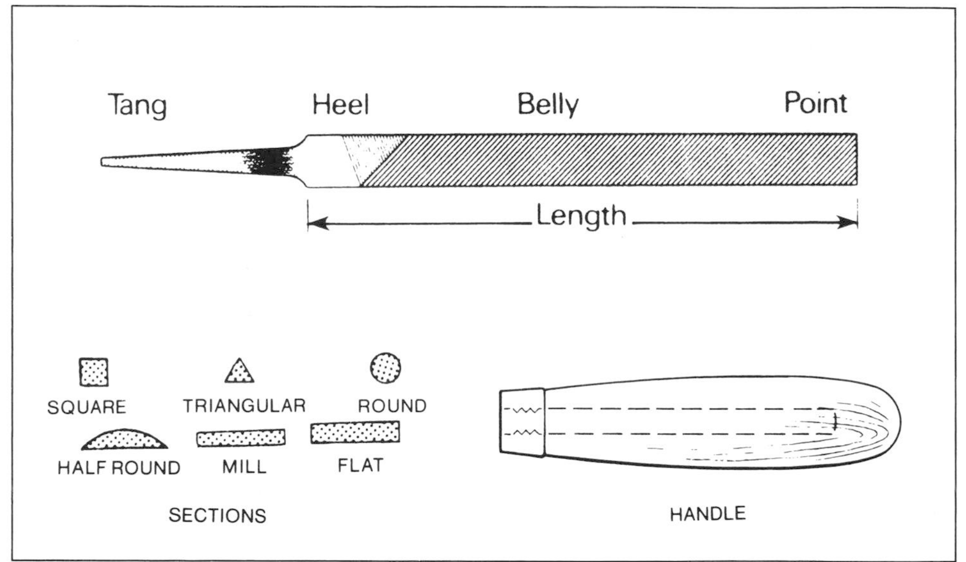

A file and its terms. Get an assortment of files. You'll find the second-cut, half-round file the most useful. Use a handle on each of your files to prevent cutting yourself.

also less expensive than the straight Aviation snips. The Wiss brand of snips is a good choice. They can be found in most hardware stores.

HAND SHEARS

Hand shears are power tools for cutting metal. Whether they are powered electrically or pneumatically they make short work of cutting even the toughest steel. Some hand shears are very heavy-duty and can cut even 12-gage sheet metal. All shears come with instructions specifying the limits of the tool's capacity.

Hand shears are used primarily to *blank out* metal—cut the general shape for the piece to be formed. You can cut freely in any direction on full sheets of metal. A band saw can cut metal too, but is limited by the depth of its throat. Hand shears are not limited, and can move easily. Light-duty shears can handle lighter metals with ease. Hand shears however, are somewhat inadequate when it comes to cutting close radii. I use shears for straight line cuts and hand snips for cutting close curves on metal.

I highly recommend the Kett model K-100 electric or pneumatic shears. The cost is relatively low for the capacity of the shears. They cut long, straight lines easily and very quickly.

POWER DRILL

A good *power drill* is another one of those "must haves" for sheet metal work. Whether you choose an electric drill or a pneumatic drill depends on which you have access to more easily: electricity or compressed air.

There are so many good electric drills on the market it is hard to decide which is best. The main thing is to get one with a 3/8-inch chuck, and that you also spend a few more dollars on some high-quality drill bits, which will save you money in the long run.

FILES

Files are used as often as any other type of hand equipment in sheet metal work. Every cut edge you make has to be smoothed out or rounded down.

Files are classified by shape, length and cut. *Length* is the distance meas-

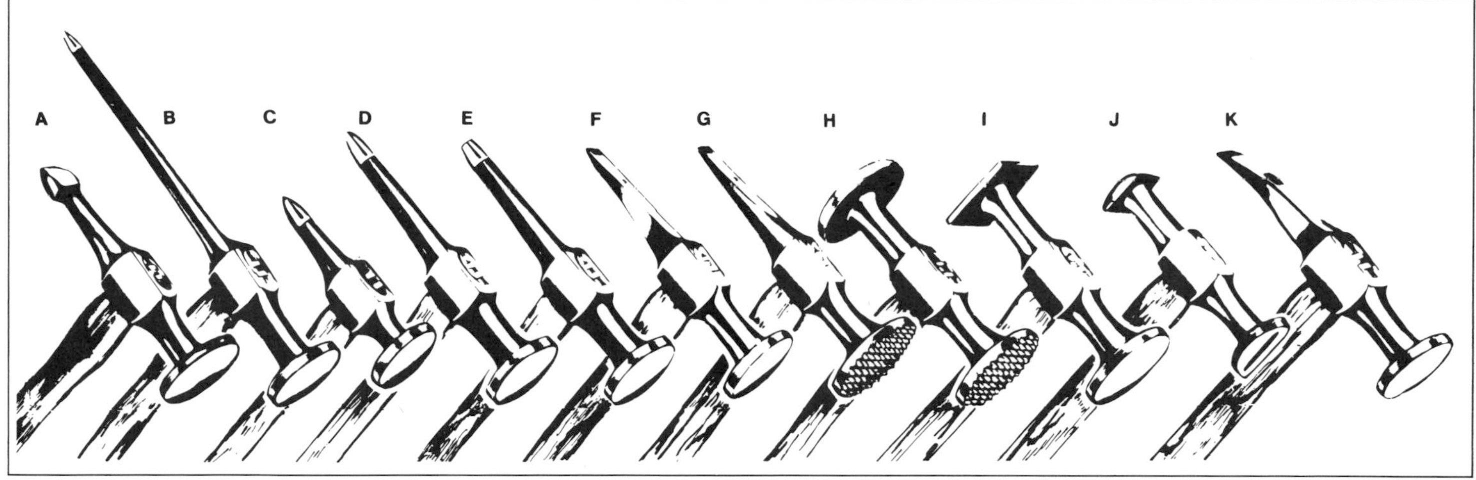

Snap-On Tools' line-up of body hammers: A: wide-nose peen; B: long picking; C: picking and dinging; D,E: spot pick; F: cross-peen; G: cross-peen shrinking; H,J: shrinking; K,L,M: bumping; N: cross-peen. Photo courtesy Snap-On Tools Corp.

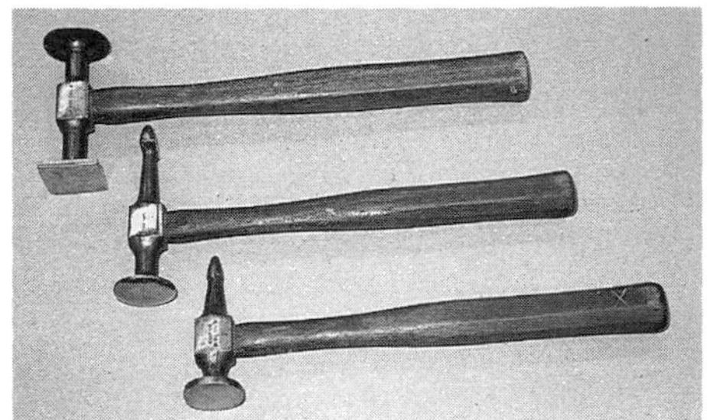

These Snap-On hammers are expensive and high quality. Some metal workers swear by them and won't use anything else.

ured from the shoulder—where the tang begins—to the tip. The *tang* is the pointed end of the file on which a wooden handle is installed. The *shape* is the shape of a file's cross-section. It may be round, half-round or flat, triangular or square. Most shapes are *tapered*, as well. The *cut* determines how much metal the file can remove at one time. A *bastard-cut* file will take off alot of metal, but it leaves a rough surface. A *second-cut* removes less metal, but leaves a finer finish. Half-round, flat and round files are the ones I find most useful. I prefer a file about 8- to 10-inches long.

Never store files together in a drawer, or pile them on top of one another in a box because they will dull one another. It is best to keep them separated, such as in a rack where it is easy to choose the right one quickly. It is important to keep wooden handles on all of them to avoid being cut by the tang. A *file card*, a brush for files with metal bristles on one side and soft bristles on the other, will keep the files clean. It is important to keep files dry, and to brush them with the file card after use.

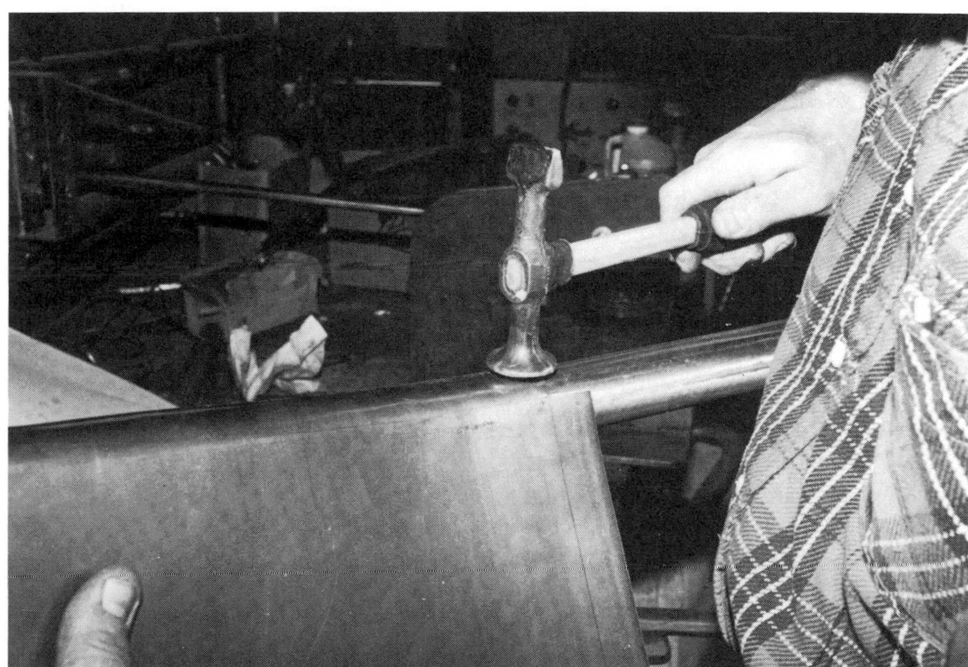

My favorite hammer is the Proto 1427 body hammer, which seems to be in my hand all the time. It's versatile, and has a replaceable fiberglass handle.

Files can be easily purchased from a hardware store, or from a tool distributor. Start with a basic set, and add more files of various shapes and sizes when needed.

SHAPING TOOLS

The shaping tools I'm going to list here are those essential to even the most basic project. They are all hand-held, and relatively inexpensive. These tools allow you to form sheet metal into just about any shape desired. There are other tools that are actually machines, large pieces of equipment to shrink, stretch, bend and shape sheets of metal. Those will be covered in the next chapter. However, whether or not you decide to invest in those items, it will be necessary for you to purchase these basic tools.

HAMMERS

Hammers are the most common tool used in sheet metal work. There are actually quite a few different types of hammers used, however a novice sheet

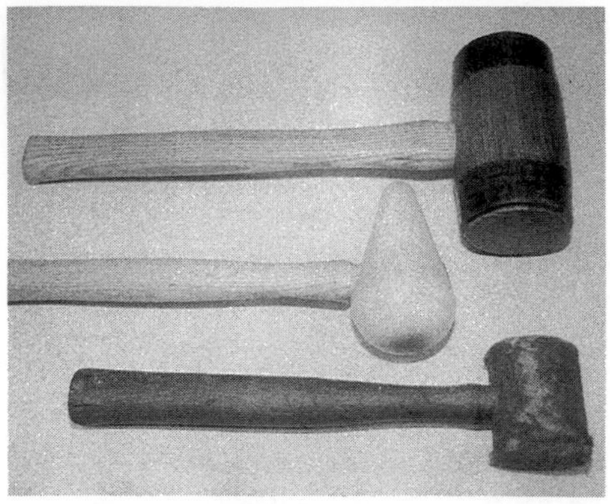

These three mallets are great for work on aluminum. When used carefully, they leave no marks on the metal. They're available at U. S. Industrial Tool and Supply Co.

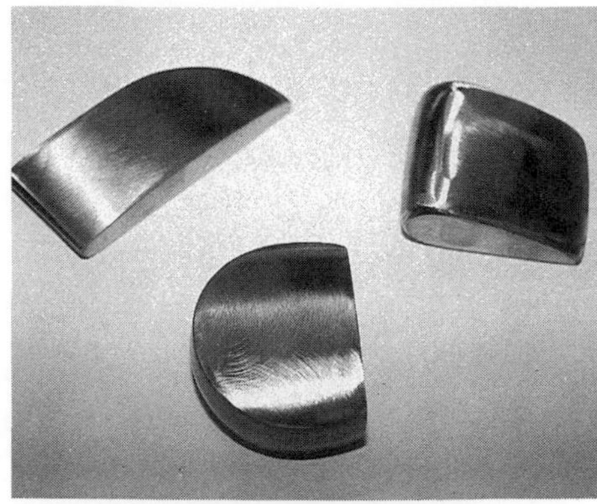

Start your collection of dollies with the toe, heel and egg dollies. The combined shapes enable you to form many curves.

metal worker should only start with three basic types. Every beginner should have a *dinging hammer*, a *shallow-domed face body hammer* and a *pick hammer*. All of these have qualities that will be constantly useful.

A basic *dinging* hammer is a body hammer with both ends of the head ending in a full flat surface. It is primarily used for flattening metal. I suggest purchasing the smaller of the two sizes commonly offered by most tool suppliers.

A *pick-hammer* style body hammer is used to bring up low spots in sheet metal. The head has two shapes: one end of the head is flat like a dinging hammer, the other end is pointed like a pick-ax. This type of hammer is available with different lengths of picks, with varying degrees of sharpness.

A *shallow-domed face* body hammer is used to form curved surfaces. I use it with the *cross-peened* end most often. The combination of the shallow-domed and cross-peened ends makes it easier to shape in a continuous curve. It is my personal favorite kind of hammer, one I use most often. I am especially fond of Proto's cross-peened / shallow-domed hammer. Snap-On and Mac Tools also make similar hammers that I recommend.

MALLETS

Mallets are used like hammers to shape metal. However, they are less likely to mar metal surfaces because they are softer. Most mallets, at least those used in metal fabrication, are made of either wood, plastic or rawhide and can be shaped into various forms and sizes on a disc or belt sander, or body grinder. This allows the mallet head to be tailored to suit a particular job at hand.

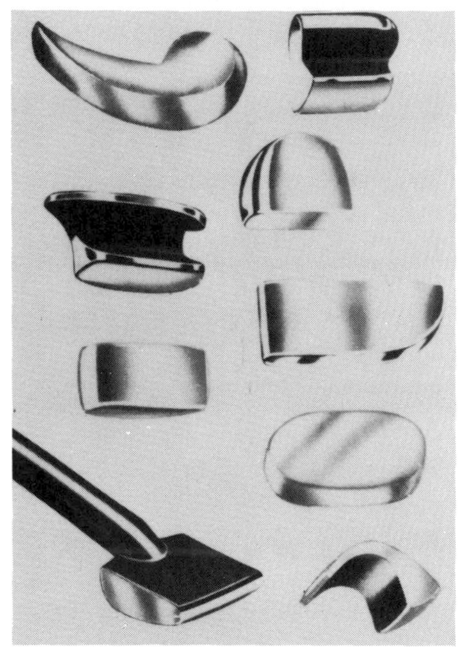

Wood mallets can be shaped or weighted with bits of steel to increase the force of the blow. They should be lightly oiled so the wood doesn't dry out and the head will stay on tight. The U. S. Industrial Tool and Supply Co. has a wide selection of high quality mallets for metal work.

The head of a *rawhide mallet* consists of tightly-rolled, rawhide leather. They tend to be more durable than wood mallets especially if they are also kept lightly oiled. Mallets of wood or rawhide are most commonly used to stretch metal and used in conjunction with shot or sand bags, which I'll talk about in a moment.

DOLLIES

Dollies, like hammers, come in a wide variety of types and sizes. They are polished shapes of cast iron or forged steel used to assist in the forming of three-dimensional shapes. Dollies are always used in combination with a hammer, to back-up the metal during forming. In a way, a dolly is like an anvil, but it is hand-held and moved in conjunction with the hammer. What they really do is offer control when hand-shaping metal.

I recommend the *toe, heel and egg dollies* as the three basic ones to start with. Each is named after the shape of the dolly. The surfaces should always be polished smooth, because any irregularity can be transferred to the metal being formed. Most dollies become marred when a hammer strikes the surface accidentally, so be careful. A good dolly is expensive, and can cost as much as $25 or more. A cheap dolly usually mars so easily it is not worth its bargain basement price.

LAYOUT TOOLS

Layout tools include dividers, squares, steel rules, punches, scribers, and measuring tapes. These tools are all essential to basic pattern design, construction and layout. Pattern-making is a critical step in sheet metal work, therefore I recommend that you invest in a good set of layout tools because you'll find that you'll be using them time and time again. To learn how to make patterns using these tools, turn to Chapter 5.

SQUARES

Squares are measuring devices consisting of two rules joined to form a right angle. They are used to measure and layout lines for cutting and bending, and to check the "trueness" of an angle. There are several commonly used squares you will need to become thoroughly familiar with.

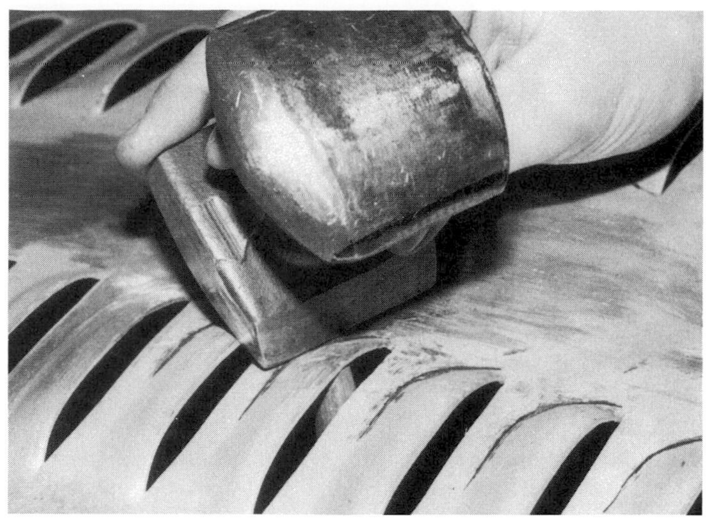

The universal dolly even has a sharp edge, which helps define edges like the base of these louvers. Add this dolly as you progress.

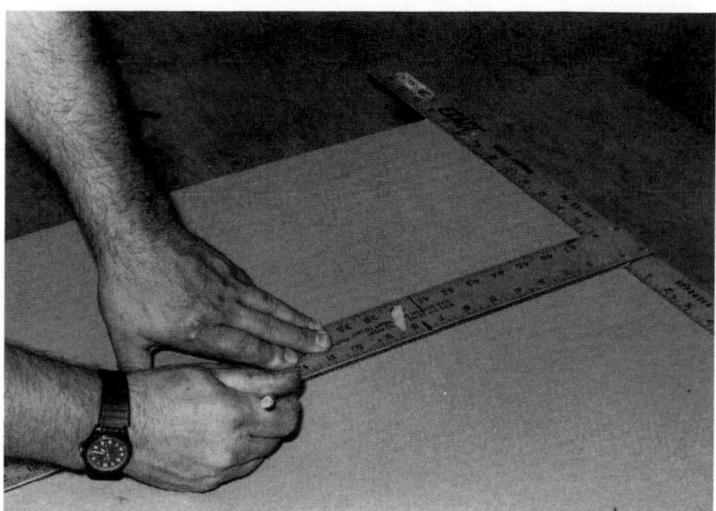

Big parts need a big square. This T-square is ideal for marking parts up to 48 inches. Be sure to hold a big T-square tightly while you're marking.

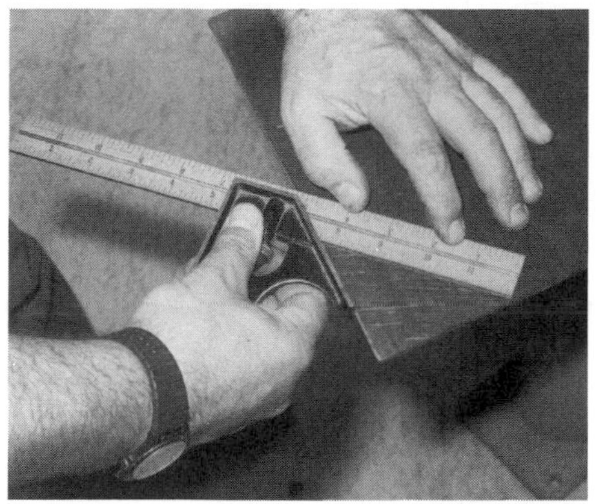

Always use a straight edge to square off from. Use the factory edge of sheet metal whenever possible. In this photo I'm using the 45-degree side of the combination square to make a triangular bracket.

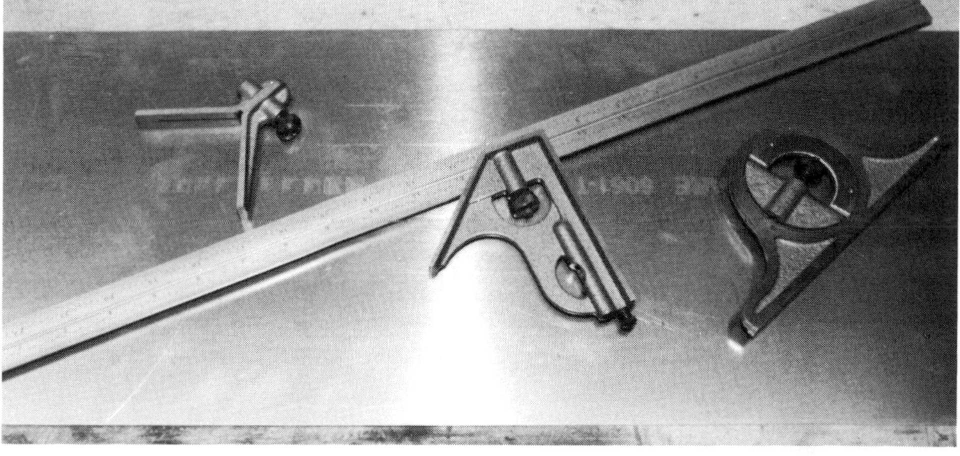

A high-quality combination square is a worthwhile addition to any tool box. Center square is at left; 45/90-degree is at center on rule; and protractor head is at right.

Carpenter's Square—The *carpenter's square* has two arms, one longer than the other. The longer arm is called the *blade*; the shorter arm is called the *tongue*. Both the tongue and blade are marked in inches and fractions-of-an-inch. Both arms can be used in layout as either a rule or as a straight edge. Carpenter squares are flat and made from one piece of metal without any joint between tongue and blade. They may be made of steel or aluminum. If aluminum, they are almost maintenance-free. If steel, you should clean and lightly oil the square periodically, depending on use.

T-Square—A *T-square* is shaped like a capital **T**. It consists of two arms at right angles to each other: a thick aluminum stock and an aluminum blade. Most T-squares are made with the stock and blade graduated in inches and fractions-of-an-inch. T-Squares come in several standard sizes: 12, 24, and 36-inches.

A *panel square* is a specific variety of T-Square that is 48-inches long and used to check long lines across a span of sheet metal stock.

In general, T-squares are very durable, able to withstand frequent use because they were originally designed for heavy construction use.

Combination Square—A *combination square* is essential if you intend to do any serious metal work, because they are very versatile. A combination square consists of a 12-inch steel scale with a *square head*, a *protractor head* and a *center head*. The heads are interchangeable, allowing you to change them around to suit the particular job you're working on at the moment. This handy feature makes layout work easier and more accurate.

The *square head* of the combination square may be moved to any position along the scale and clamped securely, or it can be removed. This square head enables you to use the combination square as a depth gage, height gage or scribing gage.

Two of the faces of the square head are at right angles to each other. A third face is at a 45-degree angle. A small level is built into the head. Some squares have a scriber housed in the head too. The level is very handy, but the scriber is almost useless for most sheet metal work because it is so short.

The center head can slide back and forth on the scale and be locked at any place. It is designed in a V-shape so the center of the 90-degree V aligns exactly along one edge of the blade.

This center head is useful when you want to locate the exact center of round metal stock. It is a specialized tool, but useful when needed.

The protractor head is also called a *bevel* protractor. It can be attached to the scale, adjusted to any position and locked at any angle. Angular graduations read from 0- to 180-degrees. A slick feature is that the degrees are marked both from left-to-right, and right-to-left, so you can easily read the scale for whatever angle measured.

The main purpose of the protractor head is to read or mark angles other than 90-degrees. A square head is more commonly used to read 90-degree angles. The protractor head is really useful when you need to measure an angle and have no idea of its degree. It will determine it for you.

The protractor head usually has a small level built in. It is a very handy feature. You can measure angles from true horizontal and true vertical.

Using Squares—Now that we've defined the different types of squares that you'll be using, let's cover a few general tips on how to use them. It is very important that *the edge of the sheet metal you intend to work from must be clean, even and free of burrs*. If the edge isn't even and clean, you can not use the square to get true measurements or angles. *Always work from an edge cut at the factory or sheared in your shop*. An edge cut in a band saw is *not* a good one to use with a square. The band saw cut is too rough and the edge may curve slightly.

When you have a good sheet metal edge to work with, be sure to place the square firmly against the edge. You must have a close, tight fit for the square to do its best work. Push down hard on the square to keep it in place, then draw or scribe the line you need.

STEEL RULES

Of all the measuring tools I use, the simplest and most common is the *steel rule*. Steel rules come in many sizes,

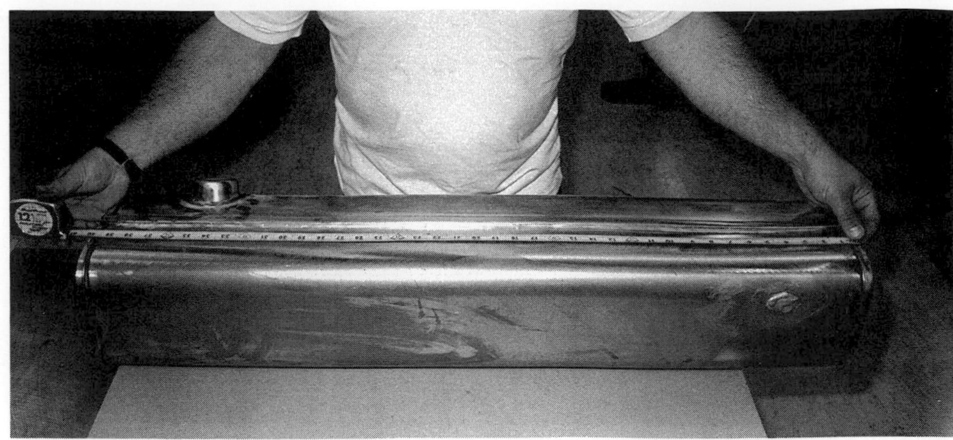

Steel tape measures are always handy. Here I'm checking the measurements of a fuel tank for a 1923 Chevrolet.

from 6-inches to 48-inches in length. The most popular sizes for metal work are 6-, 12-, 18-, 24- and 48-inch. I recommend starting out with the three basic sizes—6-, 12- and 18-inch. These sizes will suit most novice projects.

Steel rules may be flexible or rigid. Flexible rules are used to measure curved or round surfaces, like the measurement around the inside of a wheelhouse opening. Rigid rules are used to record or transfer measurements on any flat surface.

Thin rules are easier to use for accurate measurement because the division marks on the rule are closer to the metal surface. When the marks are farther away from the metal surface, you may have some trouble marking accurately. Aligning the increment with the point you're trying to measure may be tough if the rule is too thick.

There are many variations of the common rule. Generally a rule has four sets of graduations. Each edge of both sides is marked with a different graduation. One side may be graduated by 1/8-inches on one edge and by 1/16-inches on the other edge. The opposite side may have 1/32-inches on one edge and 1/64-inches on the other edge. This combination of measured increments is most useful for sheet metal work, and is very common. I recommend you choose rules with this combination.

Sometimes the increments are marked only on one side of a rule, giving only two different graduations. Generally this is less useful. I would hesitate to buy this kind of rule.

Rules should be handled carefully. Don't drop them. Don't store them with objects like files which might scratch or mar the markings. Keep rules clean and lightly oiled to prevent rust. Protect the rule surfaces from being scratched, and never let the edges get nicked. If the rule came in a case, by all means keep it stored safely in it.

Remember your measurements can only be as accurate as the rule you use to make them. If the rule becomes damaged, buy a new one. It will save you from a headache later on.

TAPE MEASURES

Steel *tape measures* are used to do *blanking measurements*, or to make general approximations and shapes on projects. The tape gives an approximate measurement, then the rules are used to refine the size. Tapes range in size from 6-feet to 100-feet in length, but the most common tapes are 12- or 24-feet in length. Most tapes are over 1/2-inch wide, and may be up to 1-inch wide. The increments are usually down to 1/32-inch. The one I like best is a 12-foot measuring tape, 3/4-inch wide. I like that size because it is comfortable to have on my belt and I rarely need to measure anything longer than twelve feet.

Steel tapes are flexible, and usually contained in a metal or plastic case. They can be either locked in position or retracted into the case by pushing a built-in button on the case. Often the case has a hook on it, so you can clip the tape onto your belt or work apron.

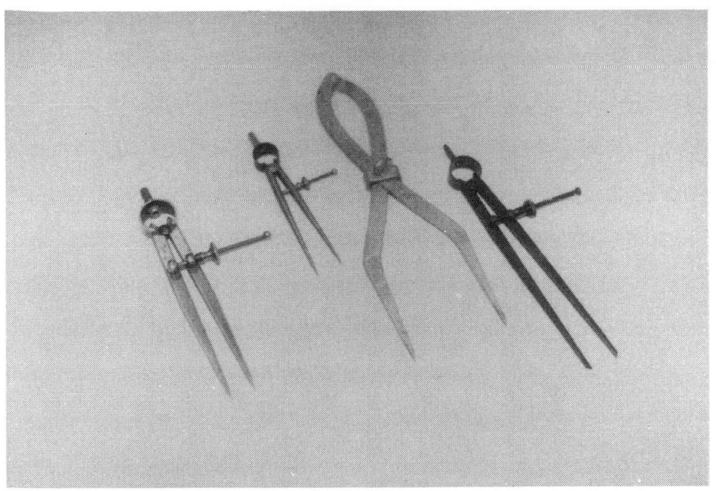

You'll need several sizes of dividers to layout various circles, or space rivet holes. These dividers have legs 3-, 4-, 6- and 8-inches long. Starrett, Malco and U. S. Industrial Tool and Supply Co. all stock them.

Scribing circles on aluminum is one use of dividers. Put one divider leg into the center punch mark, then swing the other divider leg around.

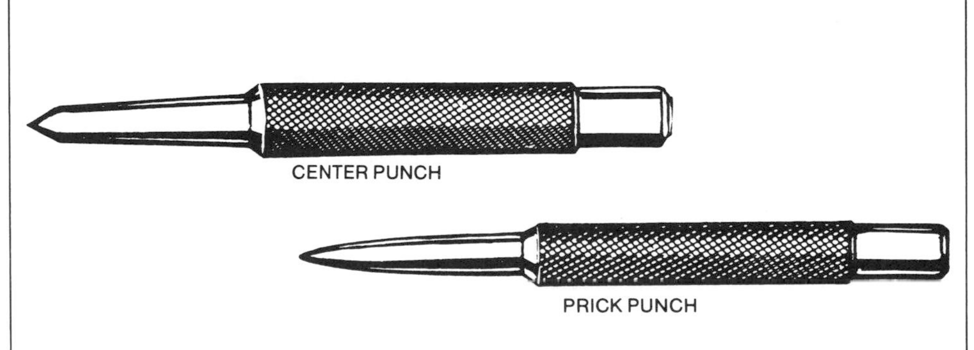

There is a difference. In addition to center punches, you'll also need a prick punch for marking metal.

DIVIDERS

Dividers are used to do all kinds of layout, but they are most commonly used to layout circles. They are a measuring tool with two movable legs, each ending in a sharp point. On some of the better dividers, one leg will be slightly longer than the other, usually by 0.050-inch, to facilitate scribing circles. The longer leg is placed into the center punch hole marking the center of your circle. Placing the slightly longer leg into the hole makes the divider legs about equal, and the shorter leg slides around to mark the circle easily and accurately. Dividers can also be used to transfer distances or divide straight or curved lines into an equal number of parts.

Dividers come in many different sizes, from 2-inches to 12-inches. The inch measurement is the length of the divider legs. I recommend you purchase two sizes: a 4-inch and an 8-inch set. Some metal work requires even larger dividers, but these two sizes are good basic tools.

These tools are very reasonably priced, and last a long time if you take care of them. Keep several sizes in your tool box. Care for them by storing them in a dry place to avoid rust. Naturally, you don't want to drop them, or mar the accurate points.

SCRIBERS

Scribers are sharply-pointed tools made of hardened steel, or sometimes carbide-tipped steel. They are usually 6-inches to 10-inches long, and used to scribe or mark lines on metal *only where it is to be cut*. Most of the other markings you make should be done only with a soft lead pencil, or by ball-point pen on the plastic covering the sheet metal itself. *Be careful scribing lines on metal*. They are permanent lines that can't be taken off without scratching the metal, which may end up affecting the quality of the finished product. Be sure that you have double-checked all of your measurements before you scribe. *The biggest failure caused by scribing occurs when you mark a bend line with a scribe*. As the piece is bent, the metal is weakened significantly at the point of stress, *especially* if the bend is in the opposite direction from the scribe line. You may be building in a place where the metal will crack and fail. Be sure you are building for strength and reliability, not future problems.

PUNCHES

Punches are another vital tool required for proper sheet metal layout. They are metal, hand-held tools used to mark holes that are used as a guide for drilling. There are several different types available.

Center Punches—*Center punches* are small, hand-held tools used to mark the dead center of a given point. They have a hardened steel tip at one end. Place the tip where a center point needs to be marked, then hammer on the opposite, blunt end of the center punch to make the mark.

A center punch is often preceded by a *prick punch*, which has a finer point to make a small mark. The point of the prick punch has a very high angle to make the mark accurate and fine. The center punch has a low angle to keep

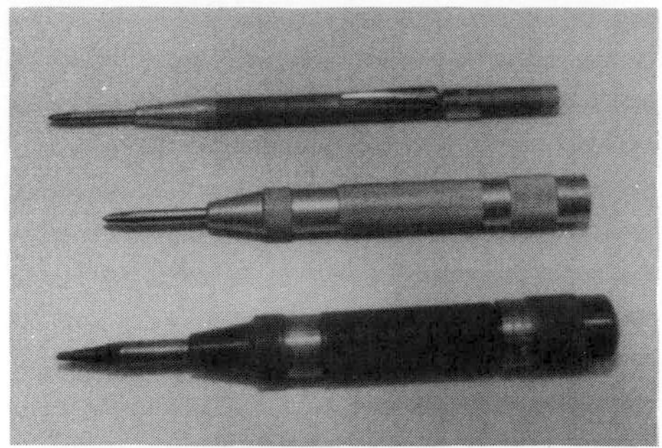

Automatic center punches ensure an accurate center punch mark. Smaller automatic center punches make smaller center marks. Larger ones make heavier marks.

the drill bit centered. The prick punch is used first, with a small hammer, to guide the center punch so the mark will be more accurate if drilling is needed. Prick punches are available in several small sizes. Center punches, on the other hand, come in many sizes, usually in a set of several.

The mark made by the center punch should be deep enough so the drill bit will start to cut, rather than wander, across the metal. If the center punch is held slightly off-angle, then the punch mark will be off-center, and the drilled hole will be inaccurate.

It is important that sheet metal be center-punched on a flat surface. The metal may be damaged if the surface is not reliably flat, or has holes in it. Holes offer no support for the blow of the center punch, and instead of a clear punch mark, there will be a dent in the metal. A dent is useless and damaging. An irregular surface under the sheet metal can cause the same problem and the same type of damage.

Automatic Center Punch—*Automatic center punches* are a great help. They make a center punch mark without using a hammer, allowing you to keep one hand free to steady the work. They are designed with a spring located inside. As the punch is pressed down, the spring is compressed up to the extent of the spring's travel (usually 1/2-inch). Then the spring is suddenly released and the point strikes the metal, giving you the mark.

The force of the blow can be adjusted as well, by turning the threaded, blunt end of the punch in or out. This allows for very accurate marking.

Automatic center punches are best

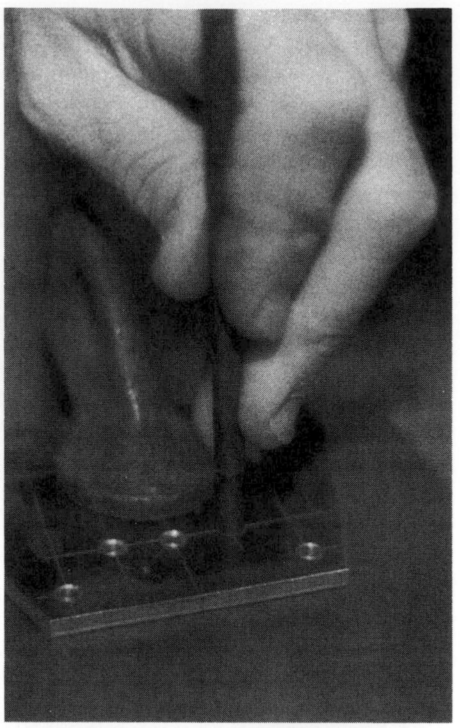

Transfer punches are a quick accurate way to transfer location information from a pattern to the metal.

for making a number of exact center punch marks for a number of holes. Since you only have to press with your hand to get the same center punch mark every time, you increase your accuracy over a regular center punch.

Automatic center punches come in three usual sizes, small, medium and large. They are very fairly priced and I strongly recommend you include them in your tool kit.

Transfer Punches—*Transfer punches* transfer a punch mark to metal at a given spot. They come in sets of several different sizes. They are used primarily when you want to reproduce a given piece which includes several holes. Sometimes they are used to transfer holes to metal from a pattern.

It is best to strike a transfer punch with a soft hammer. It is easy to mistakenly place the punch upside down, with the marking end up, and if you use a hard-faced hammer on that tip the punch is ruined. Once you've made a transfer punch, it is necessary to restrike the marks with a center punch to enlarge and deepen the mark. Transfer punch sets are an excellent idea, because they are inexpensive, durable and used often.

One big advantage of the hand lever punch: the holes it makes need no deburring. This Whitney Jensen XX can punch many different diameter holes. It's expensive, but worth the cost.

Hand-Lever Punches—A *hand-lever punch* is a strong, time-saving device that pays for itself very quickly. Making a hole usually means getting out the drill and drill bit. Sometimes a pilot hole will have to drilled first. But with a hand-lever punch, the hole can be made just by using a center mark to locate the hole. Then you use the hand-lever punch for a nice clean hole, precisely where you want it.

The two most popular hand-lever punches are the Whitney Jensen #5 Junior punch for holes up to 1/4-inch in diameter, in 1/16-inch thick steel, up to 1-1/2-inches from the edge of the sheet. If you need to punch bigger holes deeper into the sheet metal, choose the Whitney Jensen XX punch. It has a deeper throat to reach farther into the metal. It also punches holes up to 1/2-inch in diameter, in metal 1/16-inch thick, up to 3-inches from the edge of the piece.

HELPER TOOLS

If you only had the tools listed above, you could get along pretty well. There are many metal projects which would require only these tools. But if you want to make more specialized projects, then these tools are highly recommended. They will certainly make your life easier and more enjoyable, at least when it comes to working with sheet metal.

All of these tools help manage your small metal parts easier, almost as if you had an extra pair of hands. I call them "helper" tools, because they help hold metal in position to mark, cut or shape it.

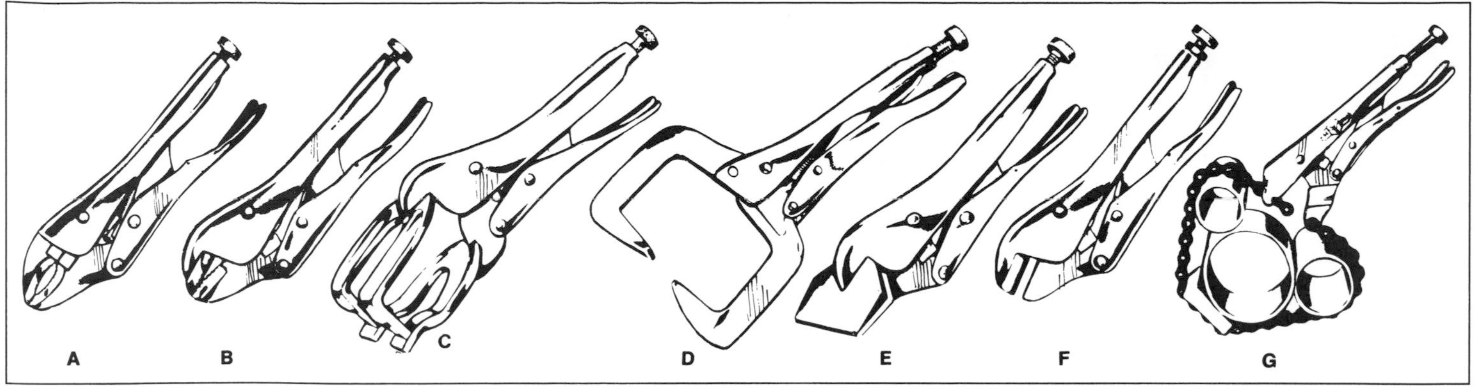

A vise-grip for nearly every clamping need. A, B: wire cutters; C: welding clamp; D: C-clamp; E: bending tool; F: hose/tubing pinch-off; G: chain wrench; Drawing courtesy of Snap-On Tools Corp.

VISE

One of the best metal-working tools around is one of the easiest to find. A good *vise* is indispensable. It holds metal securely for filing or cutting. *Never clamp a vise to a work table*. It must be bolted securely. Otherwise you will be counting on the vise to hold metal tightly when the vise itself may be loose. This will not only affect the quality of your work, but it also presents a potentially unsafe situation.

Be prepared to buy a good vise, not a cheap one. The cost of a good vise can exceed $100 or more, but this is one investment I recommend. The money will be well spent. If the tightening screw is oiled frequently, a good vise can last virtually forever.

A pair of *vise jaw covers* will keep the piece held from being marred. The covers can be removed if you should need to hold some piece very tightly and don't care about marring it.

C-CLAMPS

C-clamps are probably the most common holding tool used in sheet metal work. They come in a wide variety of sizes and shapes. They are particularly useful in securing large pieces of sheet metal for hammerforming, or for clamping two small pieces of metal together for tack-welding.

VISE GRIPS

Vise grips are like pliers, but have a special mechanism to secure the grip to metal. The jaws come in a variety of shapes and sizes, which make it possible to use vise grips to clamp items in awkward places. Some work well in a spot with a narrow, deep opening. Others are great for a wide, flat spot. I suggest that you purchase several pairs with different jaw sizes and shapes.

Our shop uses an A-frame rack of tube steel mounted on wheels to hold dozens of C-clamps. It can be rolled to wherever a project's underway, and the clamps are centrally located for everyone.

CLECOS

Clecos are another kind of temporary holding device that originated from the aircraft industry, which explains why they are commonly available from aircraft supply houses. They are used to fit and refit pieces of metal before they are permanently welded or riveted together. Their function is to hold a part in place very securely for precise fitting. The pieces can be fitted together, then taken apart again to make adjustments. Clecos, therefore, are very useful.

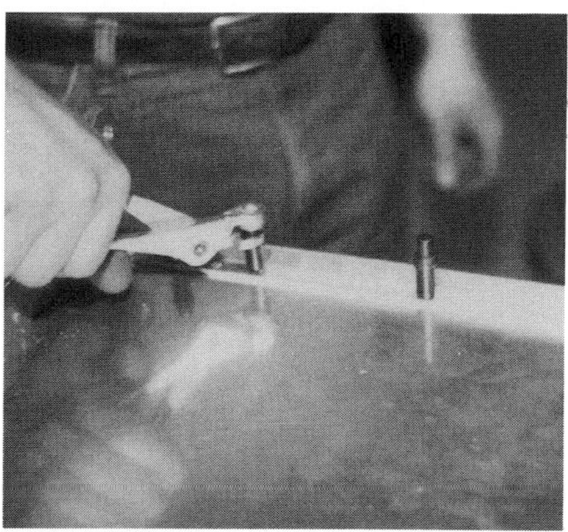

Clecos are great temporary holding devices. They are most useful in 1/8-, 5/32- and 3/16-inch sizes.

Clecos are installed and removed by special *Cleco pliers*. The jaws of a Cleco plier fit under a collar on the Cleco and over its plunger. One pair of Cleco pliers can be used for all four sizes of Clecos commonly used in sheet metal fabrication. The four most common sizes of Clecos are also color-coded. They are: 3/32-inch (silver), 1/8-inch (copper), 5/32-inch (black), and 3/16-inch (gold).

All Cleco pliers are good. The only real difference in brands may be the compactness of the plier, which affects how easily it can be used in a tight spot. Aircraft surplus stores often carry Cleco pliers and Clecos.

POP RIVET GUN

A *pop rivet gun* is one of the handiest tools in the metal working trade. It comes in several sizes and is either manually, pneumatically or hydraulically operated. This is a tool that requires thought before buying. What will be your requirements? How many

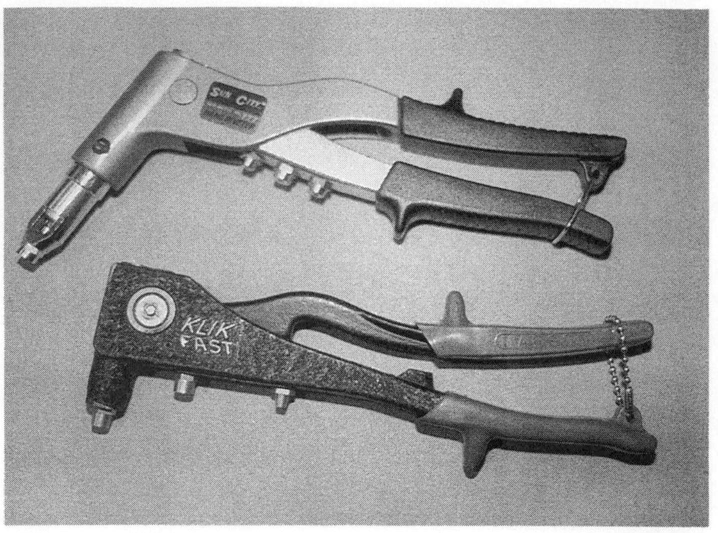

The pop rivet gun with the long nose is ideal for getting into tight places. The shorter nosed pop rivet gun is better for general use.

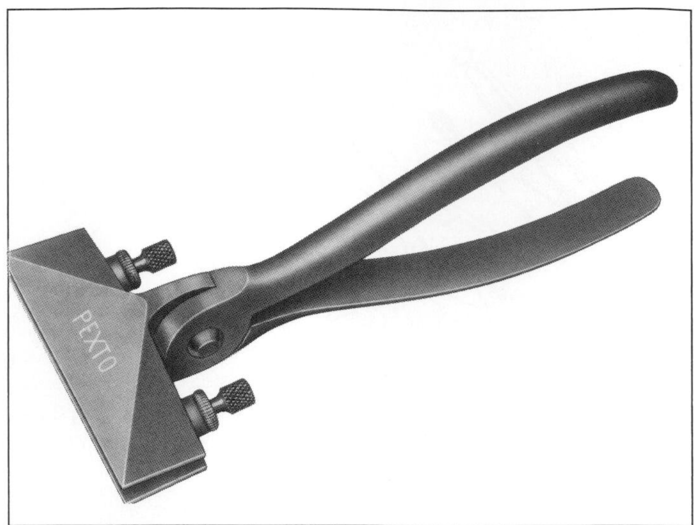

Hand seamers, both straight and offset, are used like miniature sheet metal brakes to fold metal. I use mine very frequently to bend small pieces. Pexto has a variety of jaw widths available.

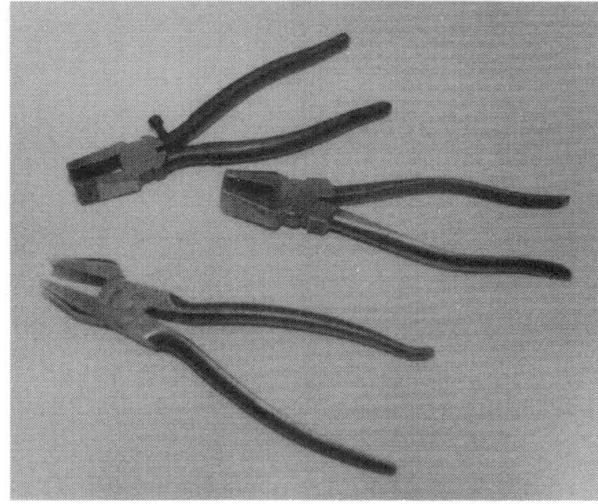

I custom grind the jaws of glass pliers to suit different needs. I leave some straight, and curve others.

One-inch wide glass pliers are ideal for bending flanges. They come with straight-jaw edges, but you can reshape them by grinding like I've done with these. A curved jaw will bend a flange around a curve. Make sure the jaws have no sharp edges or burrs.

rivets, and what size rivets will you want to use? If you want to use a few rivets of modest size, then a hand-operated rivet gun will do fine. If you intend to use alot of large rivets then a pneumatic or hydraulic gun might be the best choice for you.

SPECIALTY TOOLS

Specialty tools are a group of tools for individual jobs that aren't necessary to sheet metal fabrication on an amateur basis, but certainly help when it comes to professional work. They will make cutting or bending metal easier, and will also assist with shaping. Some are so specialized that no one will recognize what they are in your tool box. Not all of them can be purchased, either. Some are handmade for very specific applications. Adding these tools to your equipment list depends on how far you intend to go with shaping metal. If you're thinking about taking it seriously, then these items will definitely help.

HAND SEAMERS

A *hand seamer* is a small, portable version of a sheet metal brake, with jaws up to 6-inches wide, used to bend small parts. Some come with compound leverage handles to make bending by hand easier. The hand seamer, by the way, originated from the heating and air conditioning industry.

GLASS PLIERS

Glass pliers are tools used by the glazing industry to hold glass, but their

Greenlee chassis punches come in a long list of sizes. It's unlikely you'll need a size they don't make. They cut accurate, burr-free holes. All you need is a pilot hole for the bolt.

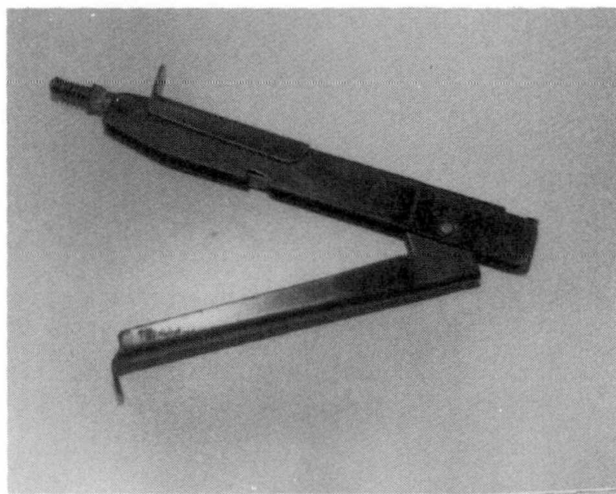

Hand notchers are a great tool to notch out small areas you can't reach with snips.

wide jaws make them an ideal, hand-held bending tool. They have no serrations on the jaws, so they don't scratch or mar metal. I took a pair of 1-inch jaw glass pliers and ground the jaws into a smooth curve. They are great for bending a flange along a curve. It is important, though, to make sure the ground jaws are *very* smooth, so you won't mar the metal as you work.

CHASSIS PUNCHES

Chassis punches (also known as knock-out punches) are simple three-piece punches used to make round holes. They can cut clean, accurate holes in steel up to 0.090-inch thick, and aluminum up to 0.125-inches thick. They must be kept oiled to reduce friction and wear on the bolt.

To use a chassis punch, a *pilot hole* that is the same diameter as the bolt of the chassis punch is drilled or punched to guide the chassis punch. Pilot holes are usually either 1/4-, 3/8- or 3/4-inch, depending upon the final size of the hole. The bolt is placed into the pilot hole. The lower die is placed on the opposite side of the metal and the screw is threaded in. The bolt is tightened to punch the hole. The result is a clean, sharp and very accurate hole. Though the chassis punch is very expensive, it saves time, which ultimately saves money. Oddly enough, chassis punches are more readily available at heating, plumbing and electrical supply houses than from suppliers of fabricating tools.

Flaring tools are homemade, simple and effective. They strengthen and beautify panels with lightening holes.

NOTCHERS

Notchers are sometimes called a *hand-nibbling tool*. They are used to cut metal by notching the edge. They can also be used to notch metal through a pilot hole, and for shaping a hole in an unusual shape, like an oval. Notchers save time filing metal away to get to a particular irregular shape. They cut equally well on steel or aluminum up to 0.050-inch thick. Notchers are also easily found at heating and plumbing supply stores, as well as some well-stocked hardware stores.

FLARING TOOLS

Flaring tools are made by hand to suit a given hole size. They are used to turn the edges of a hole into a smooth, flared edge. They are primarily used on *lightening holes* — holes made to reduce the weight of a part. The flare adds rigidity and strength to the piece which has been lightened, as well as the beautiful, finished appearance of a part professionally done. Flaring is most commonly performed in the aircraft and race car industries.

Flaring tools are manufactured on a lathe, usually made of aluminum or steel, depending on how many holes you intend to flare. Though aluminum flaring tools are fine for most applications, steel ones are recommended if you'll be flaring alot of holes.

The flaring tool is made of two pieces, an *upper and lower die*. The upper piece has a *pilot bar* that fits into the lower piece and guides the metal into the lower die as the flare is formed. The actual surfaces forming the flare must be very smooth and polished. The flaring tool will not mar the metal if it is very smooth.

T-DOLLIES

T-dollies are a special kind of dolly fabricated from two pieces of steel. In the past they had to be hand made, but

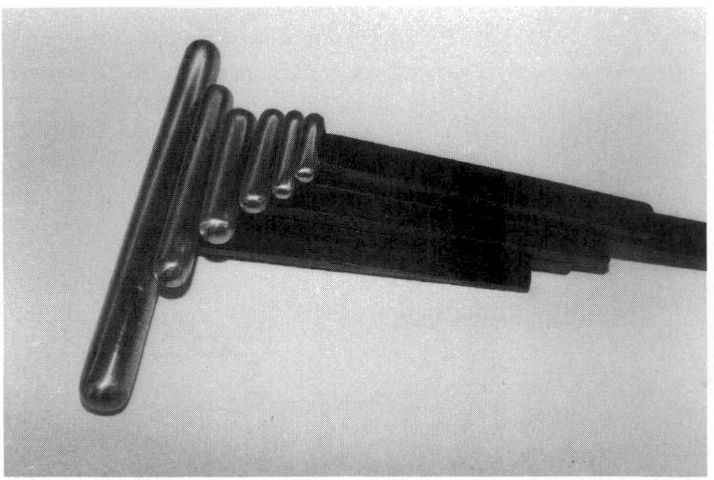

You need several sizes of T-dollies. They are sold in sets. The only place I know them to be available is U. S. Industrial Tool and Supply Co.

Slappers can move metal without marring it. These three have varying degrees of curve on the bottom. All are covered with leather.

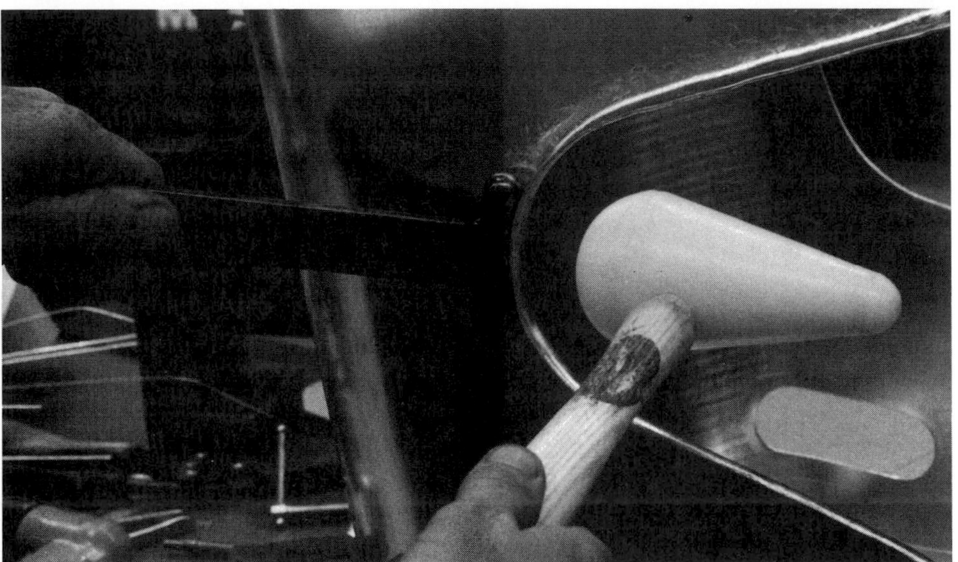

Although T-dollies are most often used in a vise, sometimes I hold one while hand-forming metal.

one manufacturer has recently begun making them. A T-dolly looks like a capital "T", and consists of the main dolly and a post. The top part of the T is made from a round bar of cold-rolled steel. Each end of the bar is ground to a hemispherical shape and polished. The post is simply used to hold the dolly. The post makes it possible to mount the T-dolly in a vise, or to hold it easily by hand. The post is usually 12-inches long, 2-inches wide, and 1/4-inch thick in diameter.

T-dollies can be used for many shaping jobs. I started out with two T-dollies and quickly discovered how useful they could be. They let me form round edges much more conveniently than any other tool.

I have made alot of different T-dollies over the years. However, U.S. Industrial Tool and Supply Co. is the lone manufacturer of them right now if you'd care to purchase them instead.

SLAPPERS

Wooden *slappers* are used like a hammer or mallet to shape metal. They're made from hard woods such as oak or rock maple. A slapper intended for use on aluminum should be covered with leather, so it won't mar the metal.

Slappers often have several curved surfaces to use in forming. There may be a flat surface for bending metal over an edge. A long, gradually-curved surface is perfect for shaping large areas over a *shot bag*. A more distinctly curved surface is used for areas requiring deep forming.

Making your own slappers is a good idea. Draw a pattern on a piece of chipboard, then transfer the pattern to a 2-inch-thick block of hard wood. Cut out the slapper shape on a band saw, then sand it smooth. Cover the slapper with leather, if you want to use it on aluminum.

Slappers are a great tool because they strike the metal in a different way than a hammer or mallet. They give a long, even blow to the metal that spreads the impact over an extended area, rather than the localized harder blow of a hammer or mallet. The curved surface, and the combination of curves and flat areas, offer control and versatility when forming metal.

CONCLUDING THOUGHT

The tools I have listed in this chapter are what I consider to be the basics of sheet metal fabrication. Most all of them are necessary to getting started. Other tools will crop up as your skill and experience levels progress. Some of those additional tools will have to be made, others are highly specialized to perform a specific task for a specific job. Of course, your budget and intent dictate the tools you will buy to get started. However, if you can assemble the tools in this chapter you will have a good solid beginning. I'll suggest additions and changes as we go along. It works best if you see how to do one thing at a time.

CHAPTER

LARGE SHEET METAL EQUIPMENT

I've already talked about hand tools, which are essential to basic sheet metal work. However, there are several large machines that you should become familiar with as well. Whether you are working weekends out of the family garage, or opening up a large business, you'll eventually come across the equipment discussed here. Most of it is relatively expensive when purchased new, however it's possible to find much of this equipment used at a bargain price. Even if you are not going to purchase any of this equipment, you should take the time to become familiar with its operation and function, simply because you never know when you may have to use it.

SHEET METAL BRAKES

The capacity of a sheet metal shop often depends on its ability to bend metal. In other words, the size or complexity of a project a shop can handle will depend on how sophisticated its bending equipment is. If all you have is a *hand seamer*, then obviously you'll be limited to very small projects that require only the most simple bending tasks.

Most all metal fabrication shops contain at least one *sheet metal brake*. It is sometimes referred to as a *leaf brake* or *metal folder*. The name "folder" is more appropriate because it accurately describes what the machine does to the metal. It folds the sheet metal along a predetermined line using mechanical force. A large variety of either simple or complicated parts can be made on a brake.

Sheet metal brakes come in many sizes and *bending capacities*. The

The finished custom pick-up box for a 1933 Willys meets specifications exactly. All the bends were made on brakes.

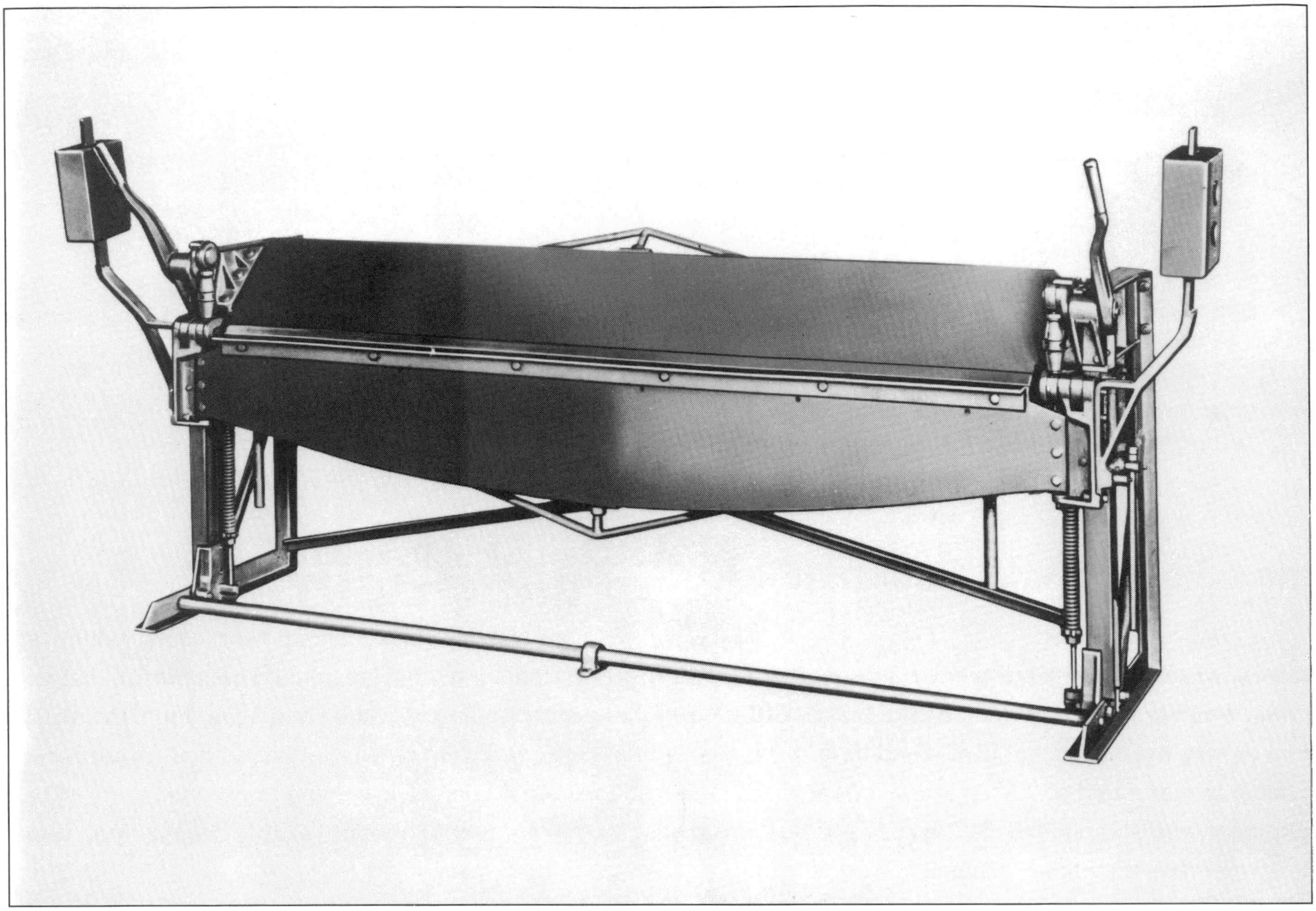

Roper-Whitney 816 sheet metal brake. I recommend it for its versatility. Photo courtesy of Roper-Whitney.

An inexpensive straight brake, like this one, is portable and can be bolted to a work bench. It has an open back which is great for bending long bends spaced closely together. Adjustable bolts on this straight brake hold the metal firmly for bending. Although it's a small brake, this will easily make crisp bends in 0.050-inch mild steel. It's a great way to go if you don't have much money, and don't anticipate having to do alot of sophisticated bending.

bending capacity refers to the maximum *gage* (thickness) and width of metal a brake can bend. The brake is usually labeled prominently with its capacity. Always check this information against the metal you're about to bend, so you don't exceed the limit. If you do, it can damage the brake.

BRAKE COMPONENTS

A sheet metal brake is made of a few parts and works simply. Essentially, a brake has an *upper beam* and a *lower beam*. The upper beam moves up-and-down so you can insert metal to be bent. When the metal is in the brake, the upper beam is clamped down very tightly on the metal. The lower beam is stationary and functions as a stabilizer to ensure accurate bending.

Another major part of the brake is the *apron*. The apron is attached to the lower beam. It is the part of the brake that actually performs bending. As the

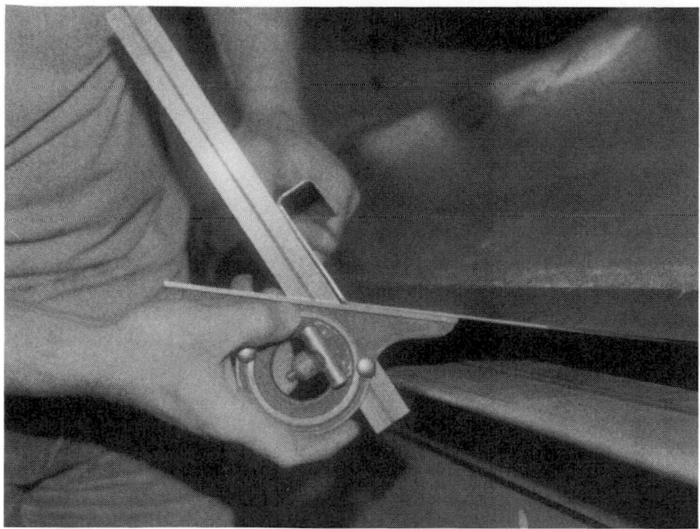

Each angle is checked for accuracy with a protractor gauge as the bends are made.

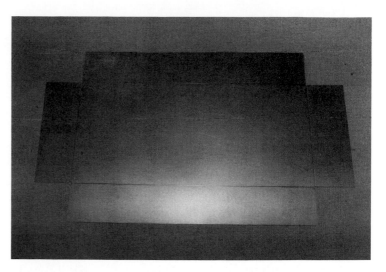

I laid out this simple pan blank to show how it is bent into a finished part on the box pan brake. The pan will be 2-in. deep x 6-in. x 12-in., made of 18 ga. mild steel.

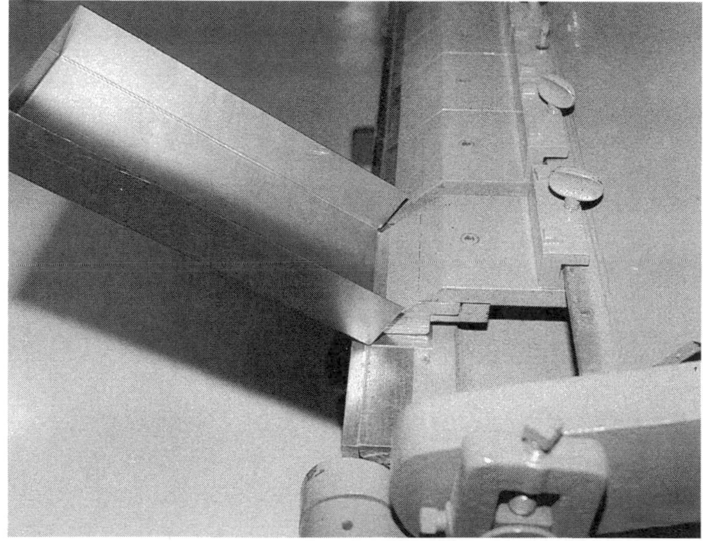

The essence of a box pan brake is the movable fingers. The beginning of the last bend shows how the pan sides clear the brake. All four sides of the pan were bent precisely 90-degrees. The beautiful fit in each corner makes final welding easier.

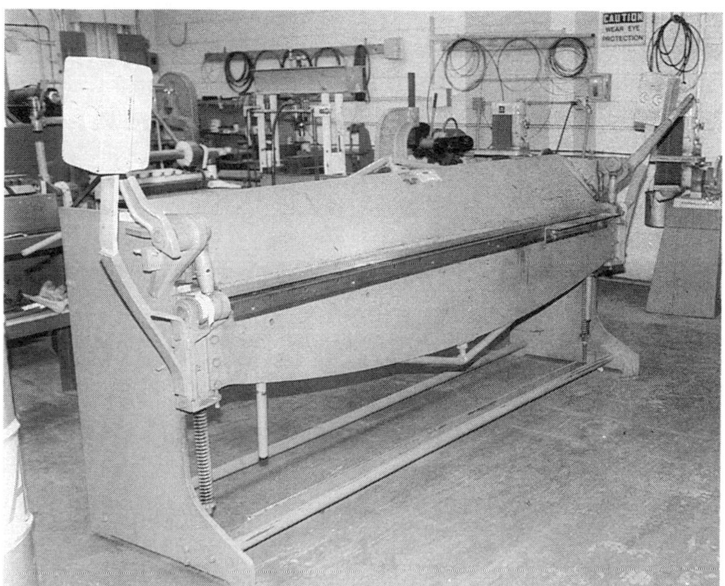

Whitney-Jensen model 816 combination brake is so useful no serious metal shop should be without one.

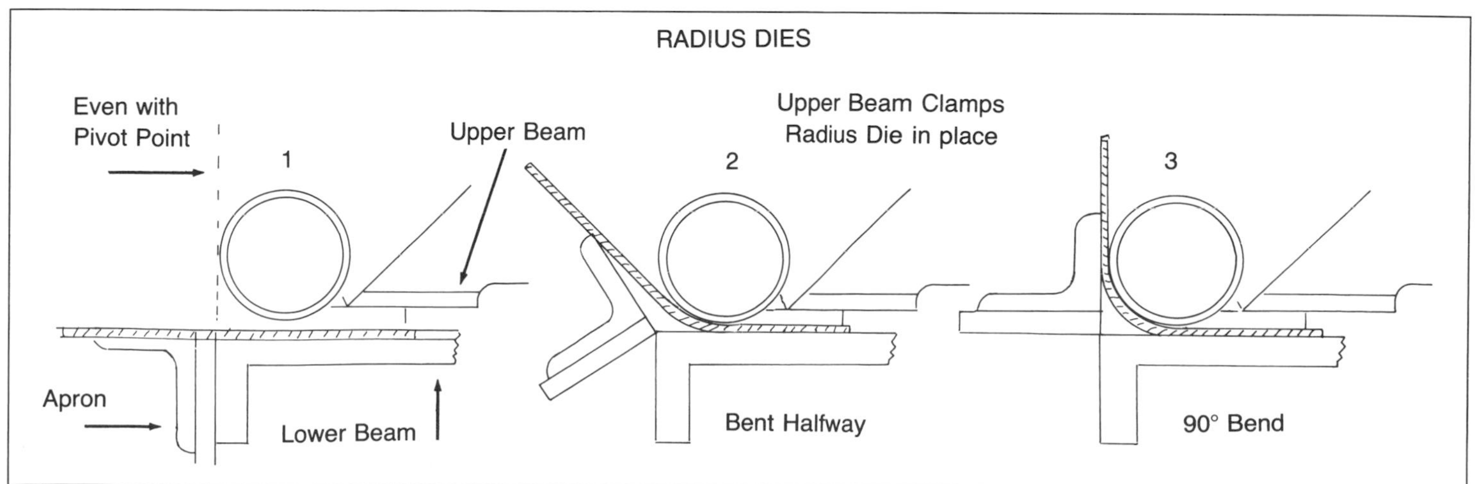

By viewing this drawing from left to right you'll see how a radius die works in a combination brake.

apron is pulled up, it connects with the metal and forces it to bend.

Most sheet metal brakes are hand-operated, though some are pneumatically-powered. Those types are usually found on a mass-production line. The brake you'll use most often has a *hand-lever* located on either side that contains a counterweight made of heavy metal. As the lever is pulled up, the counterweight lowers to make the bending motion smooth and relatively light, with little resistance.

TYPES OF BRAKES

Straight Bending Brake—This is the first of the three most common sheet metal brakes used. This brake can only make bends that are parallel to one another. Its use in a general metal shop is limited. It can bend angles and U-sections, or fold an edge completely over, but that's about it. The parts for custom cars or aircraft are generally more complicated. I only use a straight bending brake occasionally.

Box Finger Brake—is a bit more versatile than the straight brake. The *bending edge* of the upper beam—the edge against which the metal is bent—is made of adjustable parts called *fingers*. The fingers are heavy bars of metal that are bolted to the upper beam. When all are in, they form one continuous bending edge. However, the fingers can be adjusted or removed to make a wide variety of special bends, such as those required for making a box or pan shape.

The most important difference between the box finger brake and the straight bending brake is the box finger brake can make bends at angles to each other, not just parallel. This feature allows you to make a wider variety of projects with varying levels of complexity. For most applications, I'd recommend a box finger brake four feet long with a bending capacity up to 12-gage steel.

Combination Brake— The name says it all. The combination brake gives you the choice of making long straight bends or ones at different angles. Straight bends are easy to accomplish because the bending edge is continuous, not segmented into fingers to

The radius die produces beautiful round bends. This die formed a 1/2-in. radius bend in 18 ga. mild steel.

The first step in making a big curve on a brake is to mark a series of closely spaced parallel lines. Close spacing of bend lines—these are 1/8-in. apart—lets you make many very small bends which add up to one continuous long curve in the metal.

The finished smooth curve of this plenum is the result of 22 individual, equal bends.

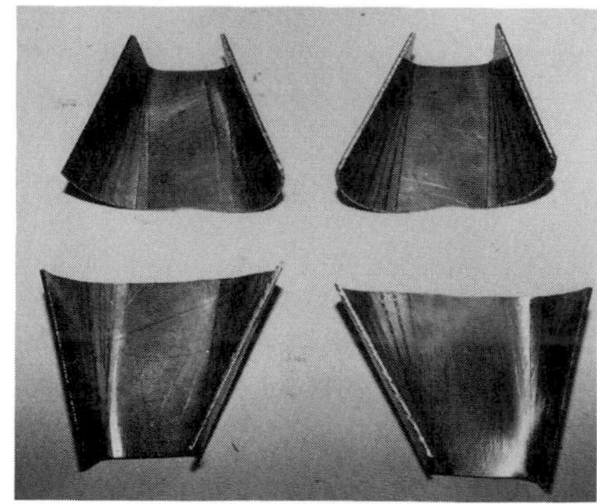

You can bend a cone shape by marking and bending lines laid out in a fan pattern. These conical parts are for a turbo exhaust collector. They are 321 stainless steel. Bending them this way was quick and easy.

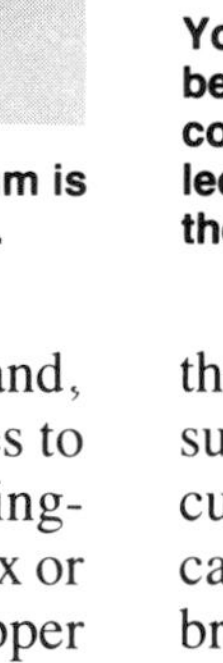

install and remove. On the other hand, if you want to make bends at angles to each other, you can bolt on some fingers on the bending edge to make box or pan shapes. In addition, the upper beam can be moved back ten inches or up four inches to utilize another feature unique to the combination brake.

The other feature is the ability to use *radius-forming dies* in the brake that can be bought or hand-made. With these dies, you can form large *radius bends* on sheet metal. The combination brake is the most versatile of the three types listed here. Special parts such as the "skin" for a race car wing, a curved dash panel or the side of a race car tub are possible with combination brakes. In fact, their use is only limited by your skill. The combination brake is best for a busy, large metal shop. I recommend one eight feet long with a bending capacity of 14-gage steel. This should handle most any metal used in sheet metal work.

MAKING BENDS

Making a bend is a simple process,

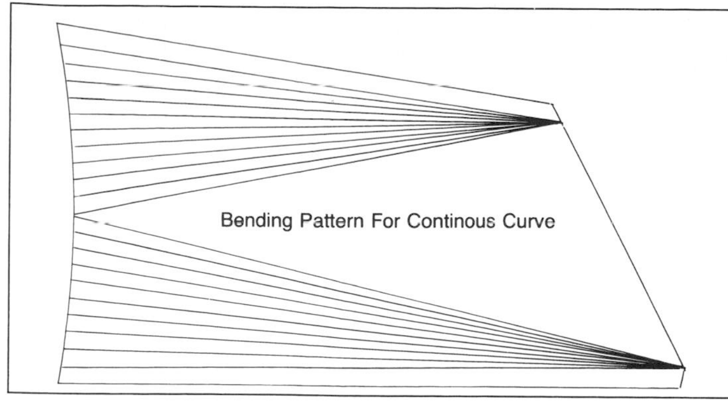

Twenty bend lines make it possible to form a round-to-square transition. I made this pattern after a trial fitting with a sheet of chipboard. This pattern will produce the bottom-half of a throttle-to-plenum connector.

Now that the bottom is bent, I'll bend a nearly identical top-half and weld it on. The throttle-to-plenum connector will then be complete.

but you must be careful. First, mark the metal piece with a pencil. A clear, thin, dark pencil line is best. **DO NOT USE A SCRIBE!** (see sidebar). After marking the sheet metal where you want to bend it take the metal to the brake. Line up the pencil line with the upper beam of the brake. Clamp the sheet metal tightly with the upper beam, then check and double-check the alignment of your marking line against the upper beam. Make sure it is correct! Pull up on the apron to do the actual bending, as far up as required to get the desired bend angle.

When you have completed the bend, release the upper beam. Remove the metal and check the degree of bend with a square or protractor to see if it's what you wanted.

MORE ON BRAKES

A brake will last for a long time if not abused. A common mistake is to set the brake improperly. The other is to bend metal too thick for the capacity of the brake. This mistake will bow the upper beam.

These are items you should consider when buying a used brake. If you're buying used equipment, take a few pieces of metal to the dealer. Ask to try out the brake. Examine the bend surface of the sample piece of metal, and check for the accuracy of angle, or to see if the brake left any scratches or marks on the metal surface. If the dealer hesitates to let you try the brake, then take that as a sign that there is something wrong with it, and look for your deal elsewhere.

The clear work area around the Pexto 52-inch power shear is a safety factor. You sure don't want to trip over something walking to the shear carrying a 40-lb. piece of metal—it's like a giant razor blade. Read the warning signs posted clearly. Photo by Michael Lutfy

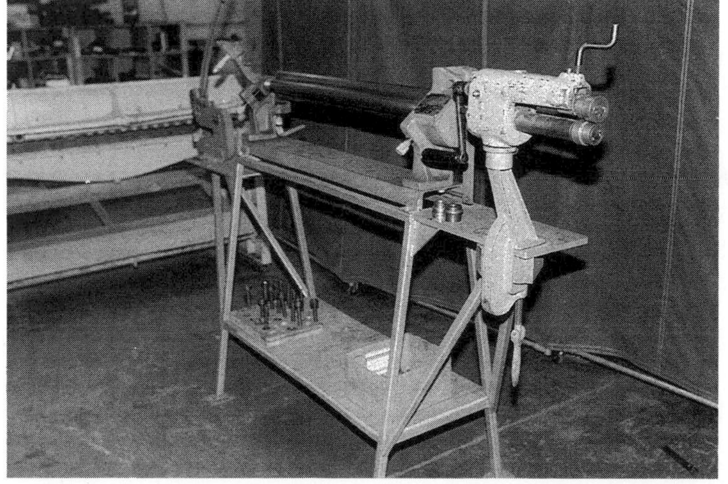

Pexto's 36-in. x 2-in. roller is the standard of the industry. It's in almost every sheet metal shop. I added a Pexto beader and a sheet metal punch to the roller stand, which created a work station and kept a work bench clear.

MAKING BENDS WITH BRAKES

1. Remember: do not use a scribe to mark a bend line. A scribe line weakens the metal and may cause a fracture, either when you bend the metal or after the part is in use. A fracture can be dangerous.
2. It is best to be conservative as you bend, because if it is not quite enough, it is always easier to bend the metal a bit more, rather than bend it back if you went too far.

A true sheet metal shop would not be complete without some kind of sheet metal brake. Look around and choose one to satisfy your needs. As you start out, a straight brake may be all you need, then as your skills and business grow, you can invest in a box or combination brake. However, remember that the straight brake will limit your bending capacity.

SHEET METAL SHEARS

A good shear is also an integral part of any sheet metal shop. Sheet metal must be cut to size constantly, and a *foot shear* or *power-squaring* shear can save alot of time. A foot shear is mechanically-powered, while a power shear is electrically-powered. Both shears are constructed of simple components and their purpose is the same: to make long, straight, precise cuts.

Like sheet metal brakes, shears come in many sizes with various cutting capacities. A foot shear is fine for cutting steel up to 1/16-inch thick. For metal thicker than that, you'll probably need a power shear.

The 52-inch shear is a good length to start with. Many people are talked into a 36-inch shear and are disappointed, because sheets of steel or aluminum generally come in 48-inch widths. I've used shears as big as 10-feet long, with the capacity to cut 1/4-inch thick steel or more. Shears are available in even larger sizes for heavy industrial use, such as in the ship building industry.

Most shears have a *side gage* located 90-degrees to the cutting edge. The side gage lets you measure the length of the cut right at the shear. It saves some time and trouble if you're just cutting down bulk sheets, and helps you make a square cut.

Some shears have a *back gage* that can be set and locked quickly. This feature is useful if you're planning on cutting many sheets of metal down to the same size. That way, you don't have to measure each sheet. The side gage is great for gaging lengths of cuts, but it is still necessary to cut on a marked line.

Another important feature about a shear is the *locator clamp*. This clamp is spring-loaded, and moves in conjunction with the blade. When the blade goes up there is room to slide the metal under the clamp. When the blade comes down the clamp holds down the sheet metal before the cutting blade comes down, to ensure the metal won't slip during the cutting.

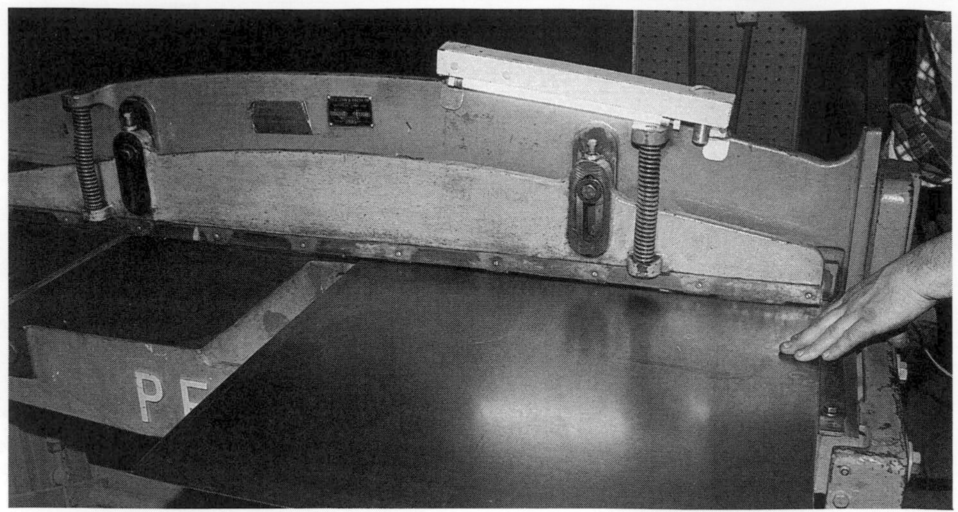

I made an 18-in. fluorescent light fixture to mount on the shear. It clearly shows the shear line, blade and where fingers are.

USING SHEARS

Using shears is as easy as marking a straight line where you want the cut, lining it up in the blade, and operating the foot pedal. It is easy, therefore, to get a bit lax when using them because "there's nothing to it." Just remember how easily they cut through steel and it isn't difficult to imagine what they can do to fingers. Most shears, however, have a plastic guard to keep fingers out of the cutting blade. Most also have a "stop" under the foot treadle to protect your foot from being pinched.

It can be difficult to see the marked cutting line on the sheet metal if the light in the shop is not very good, so it is a good idea to put a separate light over the shear blades. This will help to line up the cuts quicker. It will also help to ensure better accuracy.

There are a few "don'ts" when working with large shears. First, *don't* cut metal thicker than the capacity of the shear. Cutting metal beyond the capacity of the shear will quickly damage a relatively expensive piece of equipment. The capacity of each shear is clearly marked on the machine. Read it and follow it.

Second, *always* make sure the sheet metal is clean before you cut it. Any dirt, grit or weld slag will dull or damage the cutting blades of the shear, as will cutting across a weld, or cutting a hard rod or steel wire. A dull or damaged blade is not accurate. Blades can be removed and re-sharpened, however, at a cost that is well worth it. Just be sure to have them professionally done. Sharp, oiled and well-cared for blades will ensure accurate, clean cuts that will save hours deburring or filing.

SHEET METAL ROLLERS

Sheet metal rollers, (also commonly known as *slip rolls*), are large pieces of equipment used to curve or roll metal in a single plane. Many parts, such as a cone or wheelhouse, need to have a single, flat piece of metal transformed into a smooth, flowing, and continuous curve, or in the case of an exhaust collector, in a tight, round, tube-like shape. Sheet metal rollers are used to do the job.

The machine itself is hand-operated and comprised of a frame or base, three rollers and a hand-turned crank. The rollers are stacked with one top roller and two bottom rollers. Only the bottom rollers are adjustable.

The degree of curve is controlled by how closely the bottom rollers are set in relation to the top roller. The pressure caused by the bottom rollers

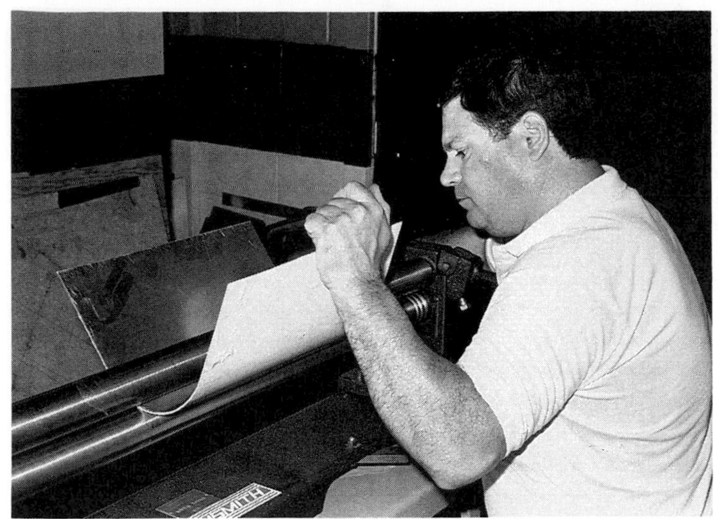
Tennsmith also makes a good 36-in. x 2-in. roller. I used it to roll the center part of an aluminum seat. Photo by Michael Lutfy.

All the parts for this aluminum seat were formed in the Tennsmith roller. The seat's now in an antique track roadster.

against the top one is what causes the metal to curve. So the tighter the rollers are set, or the less gap between them, the *tighter* the curve will be when the metal is rolled through. The more space there is between the rollers, the *less* curve there will be.

The name "slip roll" comes from the independent adjustment of the bottom rollers. When they are set equally distant from the top roller, the bend is continuous and equal. When one bottom roller is set slightly less or more than the other, the metal "slips" at one end while the other end is tightly compressed, which means one end will be curved tighter than the other one. This feature enables you to roll cones.

Rollers are available in several sizes. As the length of the slip roller increases so does the diameter of the rollers. A 36-inch machine has 2-inch diameter rollers, so the smallest curve you can roll is a 2-inch diameter curve. A 48-inch machine has 3-inch rollers, so its smallest curve is 3-inches in diameter. Experience tells me that the 36-inch rolling machine is adequate to handle most all projects in an automotive sheet metal shop.

I recommend the Pexto model 381D rolling machine or a Niagara P-402 slip roller for general shop use. A lower cost, high-quality roller is made by Tennsmith, and if cost is paramount to your business then their model SR36 is a good choice.

Of course, you can scout around for used equipment, but be sure to examine it very carefully before you part with any money. Pitted, marked or scratched rollers indicate a machine that hasn't been properly taken care of. Turn the hand-crank and listen to the gears. If they are noisy, chances are they were never greased and are badly worn. Look for missing or broken parts, such as the roller adjustment screws and the upper roller clamp.

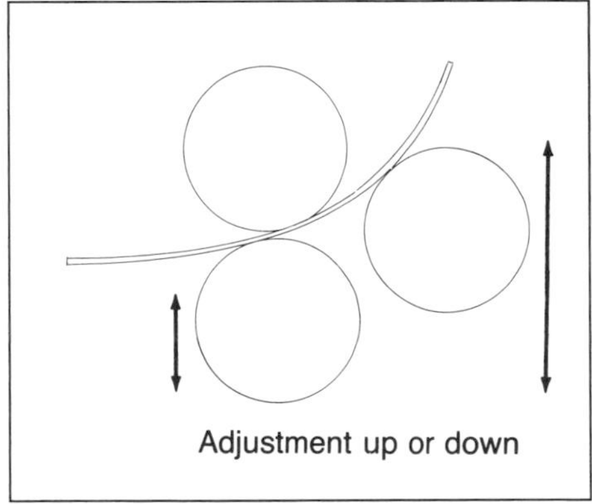

A side view of the rollers shows how the rear roller controls the amount of bend in the metal blank. The two front rollers drive the metal blank through. They must be adjusted for metal thickness.

One bit of advice to consider when purchasing a used roller: if you find one made by one of the three companies I mentioned, and the price is low because of a few broken or missing parts, then consider that they all make replacement parts for their rollers. These parts are available from their dealers who sell new machines.

New or used slip rollers, of whatever make, need care. Keep the rollers oiled and grease the gears driving them. **Never try to roll metal thicker than the roller is designed for.** It could damage the rollers. Keep the rollers from getting scratches or flaws of any kind. You can prevent damaging the rollers by being very sure any metal you attempt to roll is clean and scratch-free. An imperfection or damage on the sheet metal can be transferred to the rollers of the machine. Soft metals like aluminum are less likely to hurt the rollers, but watch out for mild steel and especially stainless steel. These metals can damage the surface of the rollers.

If you pay extra, you can purchase a stand for the slip roller. You could even make your own stand. The stand is a good idea and worth its cost. Slip rollers need to be placed so the adjustment knobs located on each end of the machine are easily accessible.

BEADERS

Beaders are hand-operated machines which form *beads*—rounded grooves or depressions in the metal. Beads are used to strengthen sheet metal panels, and as a decorative feature. A beader is powered by a hand-crank which drives an upper and lower die. The dies pinch the metal piece placed between them, forcing the metal to take the shape of the die. A crank handle above the upper die adjusts the pressure on the upper die to vary the depth of the bead.

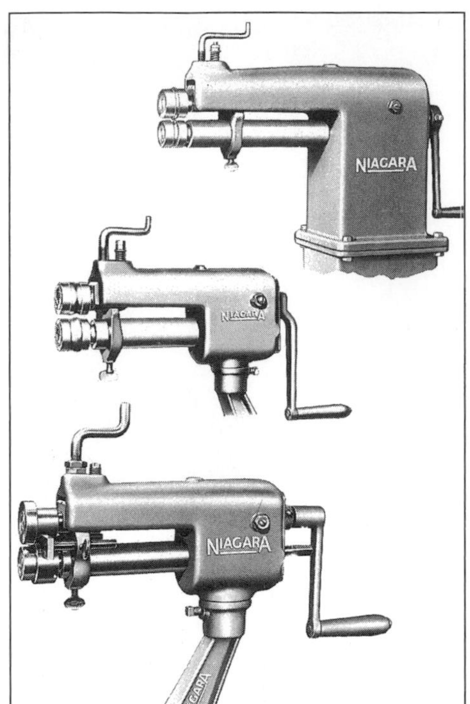

Three popular Niagara beaders are heavy duty, durable and fairly expensive. Photo courtesy Niagara Machine Tool Works.

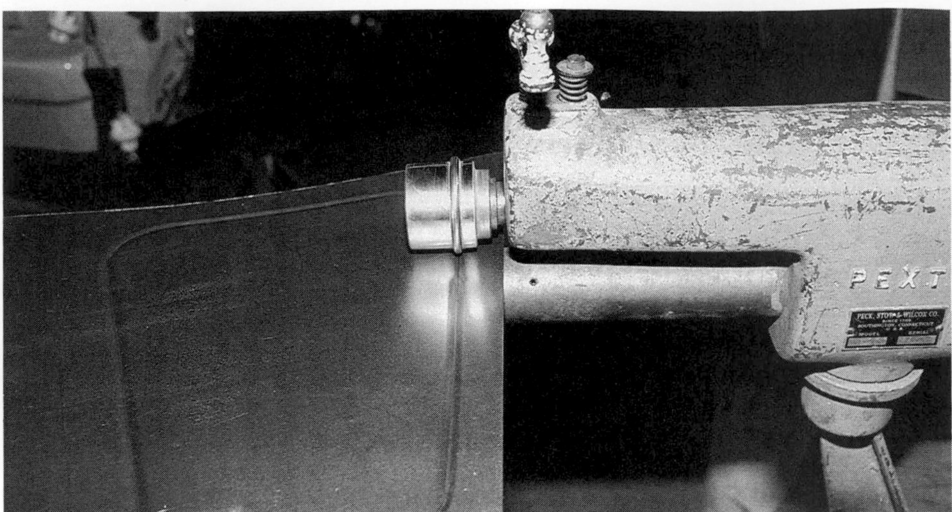

I used my Pexto 622 beader to stiffen this steel panel. I followed a pencil line to keep the beading in the right place.

A few well placed beads can enhance the looks of any metal part. Here the beads decorate—as well as strengthen—an intake manifold.

I use a Pexto model 622-E beader. It has a 7-inch throat and a large selection of different beading dies. It can form many different beads in metal up to seven inches from the edge of the metal. I use this beader most often and recommend it highly. A beader this size can handle up to 1/16-inch thick soft aluminum, or up to 0.040-inch thick mild steel.

Sometimes you can find good beaders at used equipment dealers or surplus places. Be sure to test a used beader to check for the condition of the machine and dies.

Stepping or Joggling—Beaders are most often used to roll beads in panels. However, that's by no means the only use of the machine. I use mine to *step* or *joggle* metal—to offset an edge by one metal thickness so another panel of the same thickness can be overlaid evenly.

Beaders can also be used to *indent*, which means to place a crease in a piece to be bent. This indentation *on the bend line* makes it easier to bend correctly, particularly when following a curved bend line.

Indentation requires relatively simple but special dies for the beader. These dies are homemade, and we'll discuss them more in depth in Chapter 6 on Metal Shaping, under *Bending With Beaders*.

SHRINKERS & STRETCHERS

Shrinkers and stretchers are becoming more common each year. There is a wide variety of these machines available, in many sizes and at many prices. Some are more complicated than others, but all of them do the same basic job. They mechanically shrink or stretch metal.

It is important for you to understand how these shrinkers function. They shrink or stretch metal according to the set of *jaws* mounted in the machine. The amount of shrinking or stretching is controlled by how hard you push the handle on *hand-operated* shrinkers; how hard you press the foot pedal on *foot-powered* shrinkers; or by the length of stroke set on a gage on *electrically-powered shrinkers*.

Hand-powered shrinkers/stretchers are limited by the thickness of metal they can handle, and the depth they can reach into the metal from its edge, or *throat*. They work best on either aluminum and thin steel, such as 0.063-inch aluminum or 20-gage steel.

Foot-powered shrinkers have a deeper throat and can shrink thicker metal—0.080-inch aluminum or 16-gage steel. Foot-powered shrinkers

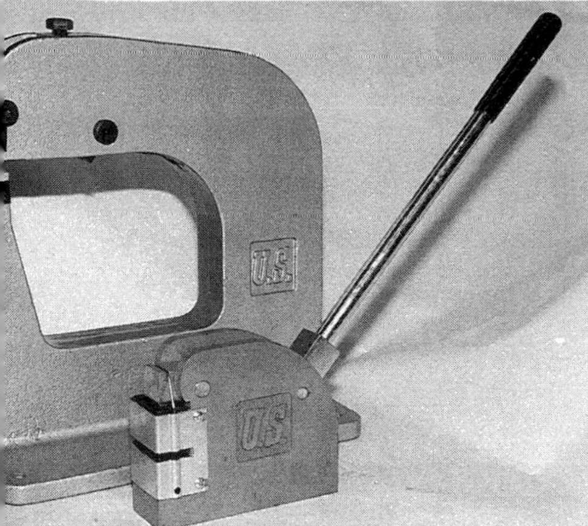

Two very popular hand shrinker/stretchers are made by U. S. Industrial Tool and Supply Co. The smaller will shrink steel and aluminum up to 1-in. deep. The larger will shrink steel and aluminum up to 10-in. deep.

also allow you to keep your hands free to control the metal.

Eckhold Kraftformers are sophisticated shrinkers/stretchers which are described in detail in Chapter 6 on *Metal Shaping*.

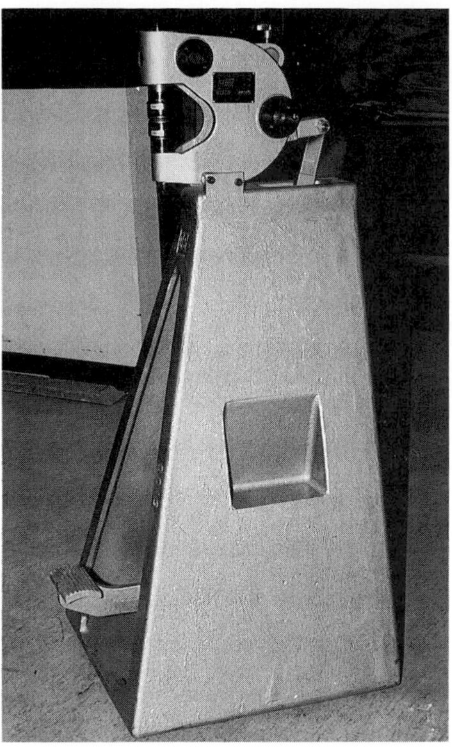

Recently U. S. Industrial Tool and Supply Co. and Eckold combined talents to produce a really great foot powered shrinker/stretcher.

BAND SAWS

Band saws are used for straight or contour cutting, depending upon the blade used. On most band saws, the work is guided into the blade by hand, as the blade is driven by the motor. The speed of the blade is controlled by adjustments on the machine. The speed at which the metal is cut, and the degree of curve are determined by how quickly the operator hand-feeds the metal through the band saw. A given blade will cut only so fast. *Never force the work into the blade to "hurry it up"*. This will only destroy the blade. Instead, choose a blade intended to cut the metal efficiently and then guide the metal into the machine slowly.

Band saws are labeled by two measurements. These measurements refer to the width of the work table from the blade to the machine, and the height from the work table to the maximum height adjustment of the blade guide. For example, a 20-inch band saw with a 10-inch height means the widest and tallest cut you could make would be 20-inches wide x 10-inches tall. The machine overall would be considerably bigger.

A band saw cuts quickly and accurately if used properly. The pieces cut by a band saw require very little clean up work. All you usually have to do is file the cut smooth, if there is any roughness. In a custom metal shop, the band saw is primarily used to cut metals of all kinds and thicknesses.

There are times when the band saw will be used to cut wood, especially when making a station buck, slappers or hammerforms. Most band saws have several cutting speeds which are used to cut different materials.

A contour cutting, multi-speed bandsaw is a must. Good ones include a blade welder. A push block saves fingers and helps control the work. Photo by Michael Lutfy

A cut-off type bandsaw is excellent for long, straight cuts. I'm using a push block and wearing safety goggles. Photo by Michael Lutfy.

BAND SAW SAFETY

The band saw can be dangerous. Injuries occur frequently because the band saw cuts at high speed and the blade is openly exposed. To avoid injury, follow these basic rules when using band saws:

1. Always have a minimum of 3 blade teeth on the piece you are cutting.
2. The blade width should decrease as the cutting radius gets smaller.
3. Keep the blade guides adjusted close to the blade.
4. Keep the movable guide post as close to the work as possible.
5. Use a wooden push block whenever possible to guide work into the blade.
6. Always wear face and eye protection. NEVER WEAR GLOVES!

As there are different speeds for different materials, so there are different blades that go with the speed settings. Wood, aluminum and steel each have a different blade and speed combination. The blade type and speed recommended for each material are listed on a chart attached to the band saw.

The *work table* on a large band saw is usually set at 90-degrees to the blade, but it can be adjusted at varying angles to suit a specialized cut.

The *blade guide* can be raised and lowered to accommodate different heights of work pieces. Always keep the blade guide as close to the work as possible. It will save wear on the blade and it helps follow the cutting line.

Some band saws have a *chip blower*. A chip blower helps keep the cutting line clear while you work. A small air compressor driven by an electric motor sends compressed air through a small, flexible metal air hose. This will keep the cutting line clear and easy to see at all times.

Blades are available two ways: in either pre-formed loops or in bulk lengths up to 100 feet. The pre-formed loop blades are easy to install and use, but they can get expensive. The bulk length is cheaper, but you must have a band saw with *blade welder and grinder* to use it.

Forming Blades— Blade loops are formed from the bulk length by cutting the length needed, butt welding it into a loop, and then grinding the weld even with the blade surface so it won't get caught in the blade guides.

Whether it's a pre-formed loop or a blade you've just welded, remember to adjust blade tension after changing blades so the blade won't slip on its drive wheels.

SANDERS

A *disc sande*r is a large, electrically-powered machine used to sand metal or wood. A disc sander comes in 12-, 15-, and 20-inch sizes. The most common size is the 12-inch. The measurement refers to the diameter of the sanding disc surface.

Design & Function— The disc sander is made of a metal housing, motor, metal rotating disc and a work table. A self-adhesive sanding disc—made of cloth coated with an abrasive—is attached to the rotating disc. The rotating disc is positioned 90-degrees to an adjustable work table. Like the work table of a band saw, it can be tilted and adjusted to the angle required.

The belt side of a combination sander has a large flat area. It's great for sanding long, straight edges. Photo by Michael Lutfy.

Larger disc sanders come with reversible motors. These machines can turn the rotating disc either clockwise or counterclockwise. That is a very big advantage. Some disc sanders even have a built-in vacuum to collect dust from grinding.

A disc sander is ideal for cleaning up rough edges from band saw cuts, or smoothing metal or wooden pieces. Different grit sanding discs are available. Each type of disc is especially suited for a particular sanding job. A 150-grit disc is good for final smoothing. A 60-grit disc is good for fast metal removal, and so on.

Discs come two ways—with or without an adhesive backing. I recommend the type with adhesive backing because they save you time and effort. The ones without adhesive backing are a bit cheaper, but must have adhesive applied before attaching them to the rotating plate of the sander. This becomes a waste of time, especially when cleaning old adhesive off of the plate every time a disc is changed.

DISC SANDER SAFETY

Use the same safety precautions with disc sanders as with a band saw. *Full face protection is essential.* A wooden push block should guide the work into the sanding surface whenever possible. Do not use gloves while sanding because they may catch in the sander and pull your hand into the abrasive surface.

COMBINATION SANDERS

A *combination disc/belt sander* is used in most metal shops. It is a compact unit saving time, space and cost. The cost it saves is the value of the time you would otherwise spend hand-filing a piece to smooth it.

A combination sander does what the name says. It combines a disc sander with a belt sander. Both sanding actions are driven by the same motor. They both have an adjustable work table. All the combination sanders I've ever come across, including ones by Rockwell, Powermatic and Wilton, have a 12-inch disc and a belt 6-inches wide. The belt is mounted on two rollers. One roller drives the belt. The other roller guides the belt and rolls as an *idler*. The idler roller is adjustable to keep the belt centered as you work.

The advantage of a belt sander is its ability to sand large surfaces. Although it can also sand edges, it really does the job on large flat surfaces. A 6x10-inch surface can be sanded safely and efficiently on the belt sander. This tool also works well on wood and plastics.

The price range for a combination sander is between $1,000 and $1,500. *Dust collectors*, those nice built-in vacuum systems, are extra.

Keep the work surface clean and dry to take care of a combination belt disc sander. Frequently clean away grit, dust and chips from your sanding. Oil the moving parts of the sander often. Keep the belt tight enough to prevent slipping in use, but not so tight it overloads the bearings. The bearings are the most likely part to wear out on a combination sander.

Rotex model 18A turret punch press has a special stand available. We bought both punch press and stand then put wheels on it to make it mobile.

Making an 0.080-in. aluminum dashboard was easier on a Rotex turret punch press. It's a fast, clean way to cut holes or blank-out large areas. Photo by Michael Lutfy.

ROTEX PUNCH

A *Rotex punch*, more properly called a *Rotex turret punch press*, is a large hand-operated punch for punching sheet metal. The punches are held in an upper and lower *turret*. The turrets are rotating carousels containing different size dies. The upper and lower punch dies must be of matching sizes before the machine can be used. Select an upper die by releasing the upper turret lock handle and rotating the turret. Then do the same with the lower turret and select the same size die as the upper one.

Each die set is numbered fractionally. A typical Rotex punch has dies ranging from 5/32- to 2-inches. They are labeled with the numbers, but also with large corresponding letters that make it easier to line them up.

USING A ROTEX PUNCH

Let's walk through the proper method to use a Rotex punch on a piece of metal. Determine where you want to punch a hole. Mark it clearly. Use a center punch to mark the metal in the dead center of your mark. Then go to the Rotex punch. Adjust the upper and lower dies to match each other and the diameter of the hole to be punched. Take the handle of the Rotex and slowly bring the handle down about halfway. There is still plenty of room to slip the metal between the dies.

Insert the metal so the upper die point falls into the center punch mark. Continue to bring the handle down slowly until the metal touches the bottom die. Now pull the handle down firmly for a clean punch. Presto! One beautiful, precise burr-free hole right where it should be.

This method of punching metal produces a beautiful, smoothly-punched hole that will not need any finishing work. That's a great help if numerous holes have to be punched in a given piece. It saves filing time.

The Rotex punch will punch accurate holes repeatedly in mild steel or aluminum. It can punch metal as tough as 10-gage mild steel, and works superbly on most aluminum sheet. Its capacity depends on the diameter of hole you want to punch. Check the owner's manual to determine the gage capacity of your punch.

The Rotex model 18A has 18 different sets of punch dies ranging 5/32- to 2-inches in diameter. The *punch throat*—the depth from the center of the die to the rear of the frame of the machine—is 18-inches deep. This offers great flexibility.

PRESS BRAKE

A *press brake* or *brake press*, is a very efficient bending machine. Press brakes are rated in two main categories: by *tonnage* and *bed width*. The *tonnage* means the amount of force, in tons, the machine can exert in bending pressure. This figure determines the thickness of metal it can bend. The *bed width* is the width of the overall machine. Bed width determines how wide or how long a bend the machine can perform.

The press brake differs from the leaf brake in that it bends metal by pressing it between two *bending dies*. The upper die rides up and down on the *rams*—hydraulic cylinders. The lower die is stationary. As the top die is lowered onto the metal it forces the metal into a V-groove in the bottom die. This quickly forms the bend. The *ram speed*, a rating of how fast the die moves up and down—is easily controlled. The tonnage is constant. The ram moves to the top of its travel very quickly. It is designed to go up quickly because it is a safe motion, moving away from the metal and hands. When lowering the die, the ram moves under precise control by the press brake operator. The ram comes down at a much slower speed for safety reasons. This allows the metal to be positioned accurately. Once the upper die touches the metal, the metal is released by the operator and the bend is formed with speed and precision.

Some press brakes of recent construction feature a special setting for the downward speed of the ram. This setting allows the ram to come down at a quicker speed until within 1-inch of the metal. Then the ram creeps very slowly down the remaining 1-inch. This feature allows the operator to check that all hands are clear before proceeding. When the ram is lowered to the point of contact with the metal, the operator pushes a foot switch to complete the bend.

The upper and lower dies are held in place with several large bolts. Some upper dies are one piece, running the full width of the press brake. Some upper dies come in several sections that can be removed or added to bend metal parts with *legs*, such as pans or boxes. Lower dies are rarely sectional. Most are one piece.

The dies are made of hardened steel. They are available in many shapes and sizes. Some bending dies come in matched sets, intended for use on a particular type of bend. The bend may have an unusual angle or material thickness. A common die combination used often is a pointed upper die used with a four-sided lower die. The big advantage to this combination is that the lower die can be unbolted and rotated to another side. The different sides of the lower die will give different bending results. One side would be for a very tight radius bend, another side would be for a larger radius bend.

A 10-ft. 400-ton press brake could bend 1/4-in. thick steel the full 10-ft. width.

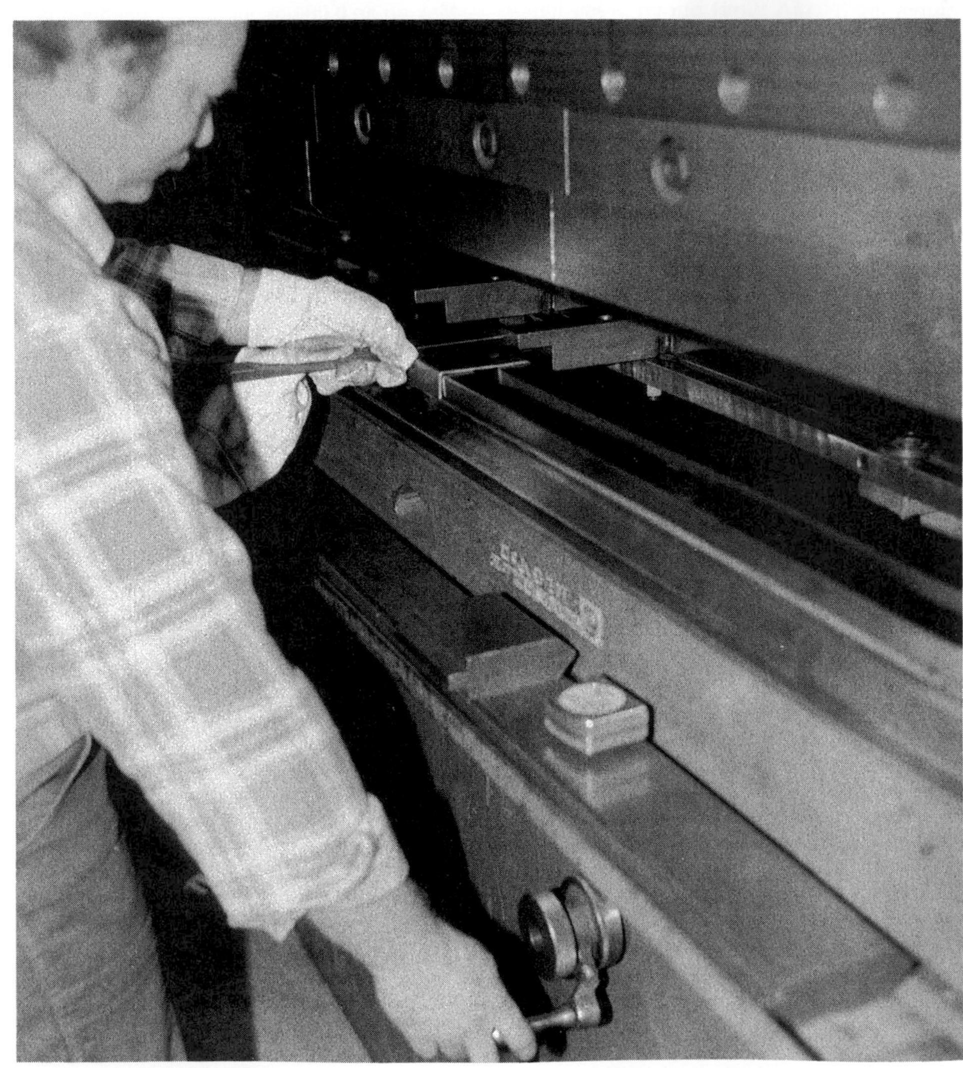

Rocky Walker turns a knob to set the back gage on the press brake. This setting is locked in place so you can repeat bends without variation.

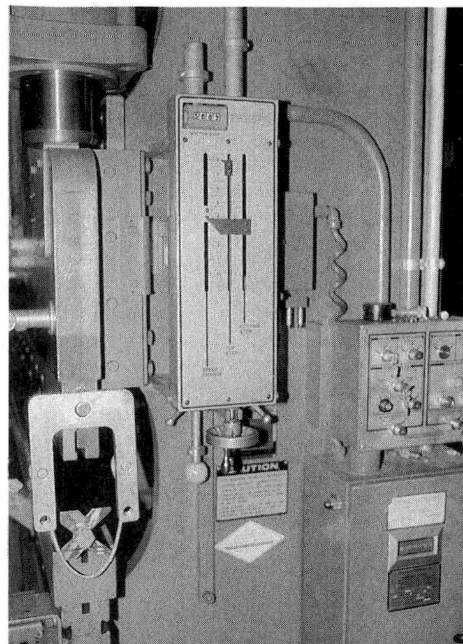

The depth of the stroke on a press brake is controlled by the setting on the control box. The more the stroke, the more bend in the metal.

USING PRESS BRAKES

The procedure for using a press brake is actually quite simple. First, mark the metal with a soft pencil to indicate the bend line. *Do not mark with a scribe because scribe lines may break when bent or when used later on.* Insert the metal into the press brake. Lower the upper die part way. Move the upper die down very slowly until the bending edge comes to rest on the marked line.

Line up the bending edge of the upper die with the marked line on the metal. When you are sure the lines are matched perfectly, push the foot control to form the bend. The amount of travel downward from the original contact point into the lower die groove determines the amount of bend.

There is a *stroke gage* on the machine which determines the distance the upper die will travel down into the lower die groove. The stroke gage also records the depth of the most recent stroke. If a greater angle of bend is needed, a deeper stroke can be dialed in. If less bend is needed, then there is going to be trouble. As with leaf brakes, it is best to bend at an angle slightly less than is required, because it is always easier to increase the bend than to decrease it.

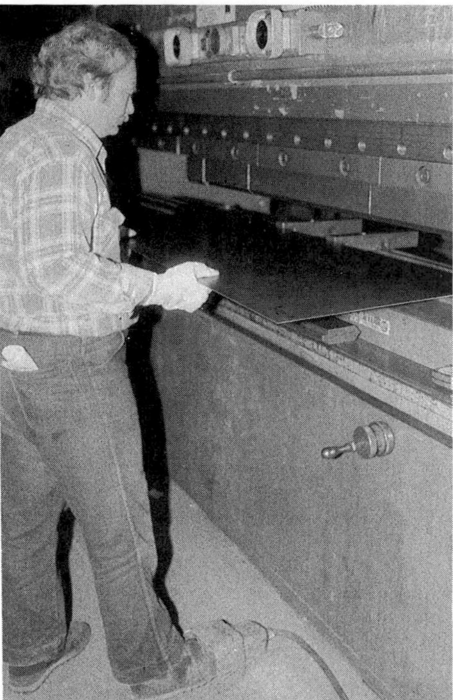

Once he's set the controls, Rocky Walker places the metal against the press brake stops and between the dies.

The ram has finished the depth of stroke and will go back up, releasing the bent metal piece.

Let's say that a 90-degree bend is desired. Set the stroke gage for a trial bend on a piece of scrap metal. Line up the metal, and make the bend. Check the bend angle with a protractor, then set the stroke gage to match the difference between the reading and the 90-degree angle. As the brake records the most recent stroke, it is easy to tell how much deeper to go the second time. Just remember to be conservative and avoid over-bending. Once you've gotten a desired bend angle on the scrap metal, duplicate it on the actual part.

A press brake also has a *back gage*, which stops the metal after it has been slipped between the dies to determine where the bend will be. The back gage is adjustable, measuring from the center of the lower die groove. If a bend 2-inches into the metal is needed, set the back gage at 2-inches. Then slide the metal until it touches the back gage. Then it is possible to make the bend without measuring and marking lines. It is a good idea, however, to make a sample part when using the back gage as a measurement to be sure the bend is occurring in the right spot. The back gage is a great help because it makes it possible to repeat bends accurately on many parts. On most press brakes the back gage is adjusted from the front by turning a crank-driven wheel.

Press brakes are becoming more sophisticated each year. They have all kinds of safety features—like press brakes you can not operate unless your hands are both clear of the dies. Some press brakes are even programmable. Spend plenty of time choosing the press brake just right for your bending needs. They are very expensive, ranging from $4,000 to $140,000 but they can be a very productive machine in a metal shop.

CONCLUDING THOUGHTS

Most of the equipment discussed in this chapter is not necessarily mandatory. The idea is to familiarize the reader with their function, and how to operate them safely. To own all of them would require an investment of several hundred thousand dollars. Some are available used at considerable savings. It all depends on what you're doing. If you're equipping a small shop, you'll probably buy one or two. However, if you're setting up a large sheet metal shop for business or to support a race car team, then you'll be interested in most of them. If you know how to use them properly, then it may be possible to use someone else's, as long as you let him know that you'll repay the favor someday. Now let's get on to our next topic, which discusses the different types of sheet metal available.

CHAPTER

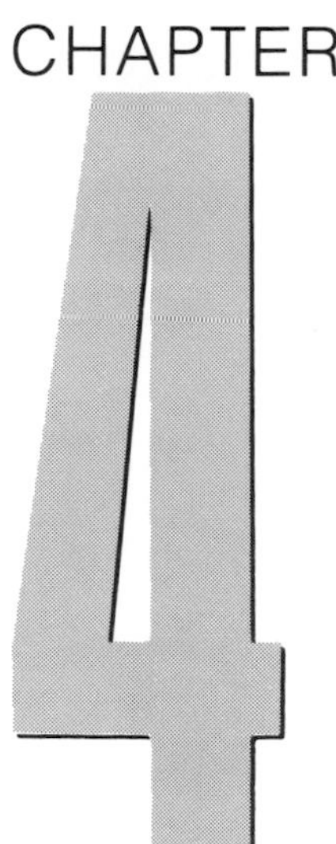

TYPES OF SHEET METAL

There are many different types of sheet metal available today. For example, one metal catalog I have lists twenty-six different types of aluminum sheet! It is very important to carefully choose the type of metal that is best, or in some cases essential, for a particular project. To do this, it is necessary to know the special qualities, strengths, weaknesses and comparative differences of the metals most commonly used.

Most metal supply houses have a salesman and an engineer who can both be helpful to suggest the type of metal best suited for a particular application.

Before contacting these people, be prepared by deciding ahead of time some of the requirements the metal must satisfy. Are you interested in saving weight without losing strength, or are you more interested in appearance? If qualities have to be compromised, which ones can be? Those are some of the questions that should be answered before calling the metal supplier.

METAL CATALOGS

Let's consider metal catalogs, because this is how metal is purchased most often. Metal catalogs are good sources of information. Some catalogs just list the size and thickness of various metals. Others are more helpful and offer detailed information about the metal, listing the quality of each metal and how it was produced.

Some catalogs are very specialized and list only steel and wire, for example. Central Steel and Wire Company has a fine catalog of just those products. Other companies list a wide variety of metal products. They list many kinds of steel and aluminum in several forms, such as sheet, rod and bar stock. The Earle M. Jorgensen Co.

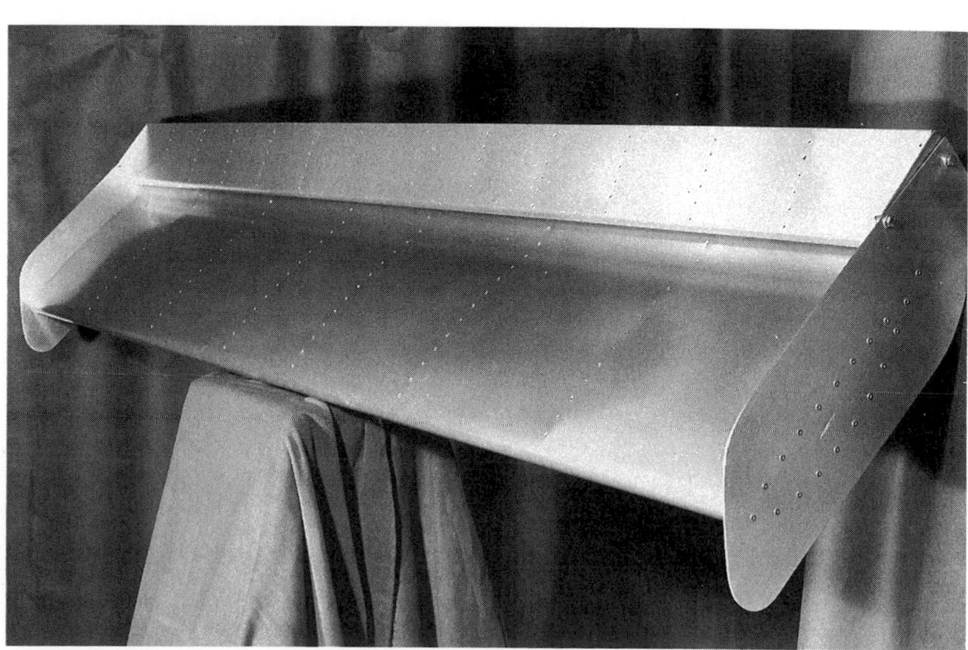

I used 2024 T4 for the wing skin, 6061 T6 for the end plates and 3003-H14 aluminum for the wing ribs. The final wing was mounted on an IMSA GTP car.

is a very large metal supplier with stores in nineteen states with a comprehensive catalog. It includes reference material that discusses all metals and their various qualities.

The Jorgensen Co. catalog, for example, is arranged with a tab index making it easy to find the section of the catalog listing the metal you are interested in. I use the tabs to get to the general section I want. Then I read their description of each alloy and see whether one or another is better for my purposes. When I have chosen a given alloy, I can then look at the listing of sizes and thicknesses available. I choose a size and thickness I think will work best for me. But it is possible I won't be able to tell which alloy is the best to use. That's when I call the metal salesman or engineer.

Be Prepared—When talking to the salesman or inhouse engineer, be sure to give them enough information in order for them to help you most effectively. Explain what you are making, and whether or not the metal will be welded, or *anodized*, which is a process where metal is coated electrolytically with a protective or colorative oxide. These are important factors. Also be prepared to tell them if the metal needs to be bent at sharp angles, and whether or not it will be exposed to weather. With this information, the metal salesman or engineer can determine which alloy and which thickness will be best suited for the job.

Ordering Sheet Metal—Assuming the type and quantity of metal has been selected, find out next what the charges are for crating the metal for shipment and for delivery. These charges should be included in the price. Sheet aluminum can be ordered with a protective paper coating, which prevents it from being scratched during shipment and can also prevent scratches while the aluminum is being bent. Charges for this vary between supply houses, so be sure to ask. *Always specify scratch-free aluminum when ordering.* Metal warehouse workers have not been above sending damaged metal if specific instructions haven't been made. If your metal is scratched, marred or dented in any way, send it back. Another tip is to check out the sliding scale of cost when buying in bulk. If buying a little more than needed for one project means hitting a price break, then it is only sensible to get a little more metal. Another tactic is to order with friends to get the price break, such as with members of a car club.

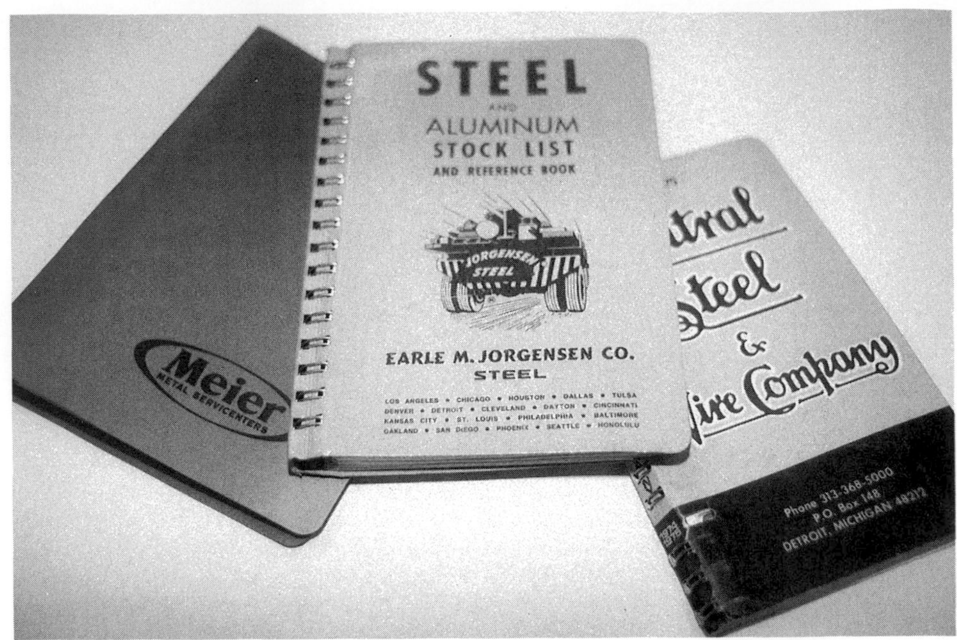

I got these catalogs by calling several local metal suppliers. They're more than glad to send them along to facilitate your order.

ORDERING TIPS

1. Always specify fresh, scratch-free material. If you don't you may be sending it back.
2. Check cost of additional protective coating. If material is aluminum, and it's for something that will be seen and possibly judged such as an engine compartment or interior, then the cost will be worth it. It is generally a good idea to order aluminum with the covering.
3. Check for price breaks that occur with bulk purchases. Sometimes you may be able to team up with friends to save some money.
4. Always be prepared with all necessary information about the project and your requirements before contacting metal salesmen.

ALUMINUM

Though there are many different types of aluminum, there are four types that are commonly used for sheet metal fabrication that will fulfill most any requirement. I'll describe the strengths and qualities of these four, as well as their various uses. There are also some special precautions and requirements to consider when using these aluminums, which I'll also be covering. First, let's consider how aluminum is labeled and coded.

CODING

Aluminum alloys are identified by a number code which indicates what element was combined with aluminum in the manufacturing process. The first number in the four-digit code indicates the major element added. I've arranged them in the following manner to make it easier to identify them.

lXXX—Designates almost pure aluminum.
2XXX—A combination of aluminum and copper.
3XXX—Is aluminum alloyed with manganese.
4XXX—Is aluminum alloyed with silicon.
5XXX—Indicates an alloy of aluminum and magnesium.

1100 H14 aluminum is a favorite of English panel beaters. This is an early '30s front fender on a Bentley.

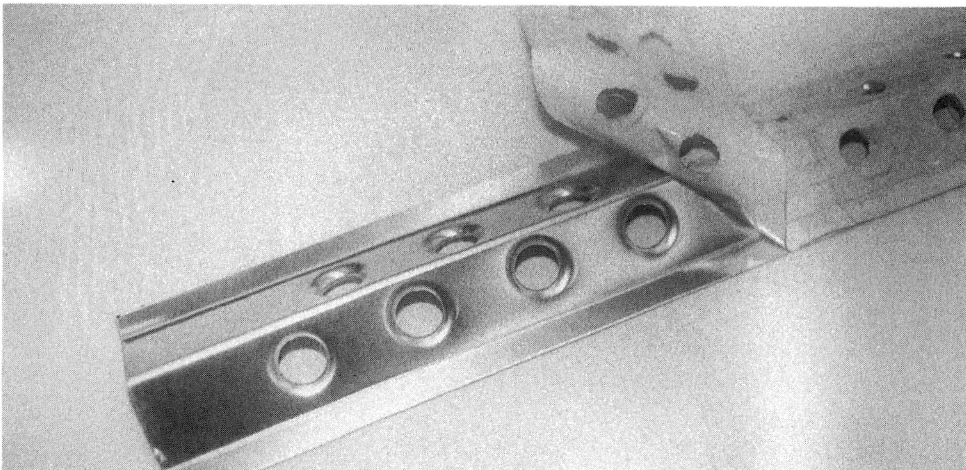

I ordered 3003-H14 aluminum, 0.063-in. thick, with a plastic adhesive cover to make this seat back brace. I actually fabricated the whole project before removing the covering so that the final part has no scratches or mars.

6XXX—Alloys containing magnesium, silicon and aluminum.

7XXX—Designate alloys containing zinc.

There are many other types of aluminum alloys, but four of these; 1,2,3, and 6 series alloys, are the ones most often used in automotive sheet metal construction.

Hardness—In addition to alloy codes, aluminum is also coded for the *temper* or manufacturing process of the given alloy. These letters serve as an indication of the *hardness* of the metal. The following are listed from softest to hardest.

F—indicates as *fabricated*, no additional processing was used to harden the metal.

0—means *annealed*, or softened by heating and cooling.

H—means *strain-hardened* by cold-working.

T—means *heat-treated* in a copper solution, sometimes *cold-worked*.

As a final identifying code, there are numbers added after the temper letters "H" and "T" to tell how much harder the metal is. For some alloys there may be several numbers showing a variety of hardness, for other alloys there may be fewer numbers. However, the *higher the number the harder and stronger the metal will be*.

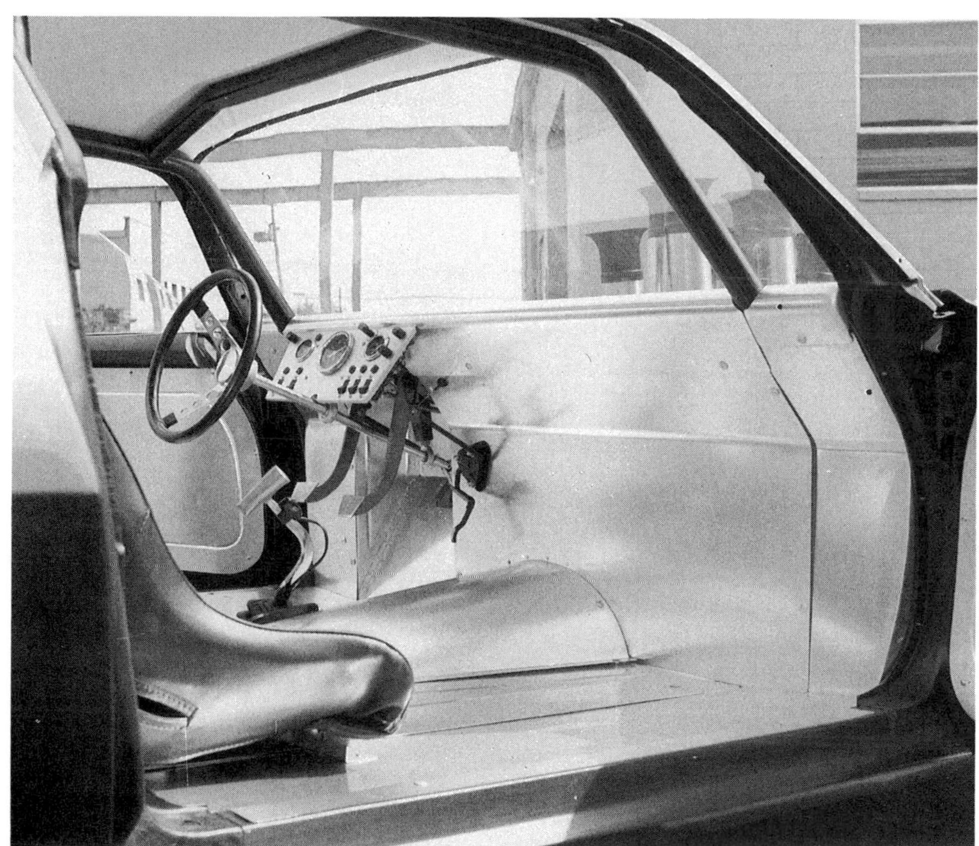

Virtually all aluminum interiors are made from 3003 series aluminum, in various thicknesses. It can be worked easily and it anodizes well.

For example, the code "3003-H14 aluminum" is an alloy manufactured with manganese added to the aluminum (as indicated by letter 3), is strain-hardened by cold-working during manufacture (as indicated by letter H), and is about halfway on the numerical hardness scale for this alloy (as shown by number 14, from scale on alloy chart in catalog). It is described by manufacturers as "very suitable for most forming operations."

Each aluminum alloy has its own

strength and forming characteristics. Let's examine the four types of aluminum sheet most commonly used.

1100 SERIES ALUMINUM

The 1100 series is commercially-pure aluminum, manufactured without the addition of any other element. It is soft and can be bent or shaped easily. This makes it ideal for projects that involve hammerforming. Another good quality of this aluminum is that it *work-hardens* more slowly than other alloys. *Work-hardening* is a term used to describe how metal becomes harder and stronger as it is being *cold-worked*—formed while cold. When an alloy work-hardens slowly it means that it can be shaped, or worked with, for a longer period of time before it becomes too hard.

The 1100 series of aluminum is also the most favorable aluminum for any type of welding. Smooth and sound welds can be made with a minimum of trouble. It resists corrosion from weather and can also be *anodized* easily.

1100 comes in five levels of hardness: 0, H12, H14, H16 and H18."O" hardness is dead soft, while "H18" is as hard as this particular aluminum is manufactured.

This is the series of aluminum most often recommended for parts that require metal spinning or intricate hammerforming.

I made the body and all other aluminum parts for my own roadster from 3003-H14, 0.063-in. thick. It shapes and welds easily and holds up to road use.

Race car wing skins like this one are frequently made of 6061 T4 0.024-in. aluminum.

Indy car tubs of the '70s used 6061 T4 and 6061 T6 aluminum throughout. The rules at the time specified 0.063-in. on all outside skins.

2024 SERIES ALUMINUM

2024 is often the alloy of choice for structures or parts where a good strength-to-weight ratio is desired, because of its strength and fatigue-resistant qualities. It is also lightweight and easy to work with.

This aluminum comes in four hardness levels: 0, T3, T4 and T361. It is an alloy made of aluminum mixed with a small amount of copper during the manufacturing process.

This alloy is most commonly used for aircraft structural components, such as wing parts or assemblies, because it offers high strength but is relatively lightweight. Two drawbacks to using this metal however, are that the bends must have large radii and the pieces must be riveted rather than welded together. It is the metal I choose, in T3 hardness, to make wing skins for racing cars, where a high strength-to-weight ratio is needed.

3003 SERIES ALUMINUM

3003 series aluminum is the most widely-used series of all aluminum alloys. It is made by adding manganese to commercially pure aluminum. This makes it stronger by nearly 20-percent over the 1100 series aluminum, yet it has all of the 1100's shaping qualities. It too resists corrosion, can be worked and hammerformed easily, and takes to all forms of welding with no problem.

It also is available in five hardnesses: 0, H12, H14, H16 and H18. The H14 hardness is the most popular—halfway on the hardness range. It is easy to work and not too hard to bend. At the same time, it is hard enough to give strength to a construction if needed.

Because of these properties, 3003-H14 is the aluminum alloy most often selected for bodywork on cars. It resists cracking from vibration or use. Gas, oil or water tanks are frequently made from this alloy in this hardness. Aluminum car interiors are another common application of 3003-H14 because it also anodizes well.

6061 SERIES ALUMINUM

This series of aluminum is an alloy of pure aluminum with small amounts of manganese or magnesium/silicon added during manufacture. It is the

1020 mild steel makes good, tough brackets which are easy to form and weld.

Other uses for 1020 mild steel are fuel cell containers, floors and firewalls.

least expensive and most versatile of the aluminums described here.

It offers a wide range of mechanical properties, such as good corrosion-resistance, brazing and welding. It can be fabricated by using most techniques. If it is *annealed*, it can also be formed easily.

6061 comes in three hardnesses: 0, T4 and T6. It can be used for a wide variety of projects. T6 is the hardness used most of the time for riveted structural constructions, like a monocoque chassis on a race car.

STEEL

Although there are a great many different types of steel only three are generally used in sheet metal fabrication. These three can satisfy the requirements of just about any project.

CODING

Like aluminum, steel is coded by numbers and letters to tell you how it is alloyed and how it was manufactured. The four-digit code tells you the percent carbon content, in *hundredths of one percent*, and the alloying elements. Letters are sometimes added to the number code to describe a process used during manufacture. For example, *1020 steel* is a basic carbon steel with 0.20-percent carbon and traces of other elements. Alloy steel, such as *4130*, is iron with 0.30-percent carbon and with a complex mixture of other elements, including *molybdenum*, its major element.

Here are the code numbers for common steels:

IXXX—is carbon steel.
2XXX—is steel with nickel.
3XXX—is steel with nickel chromium.
40XX & **44XX**—is steel with molybdenum.
41XX—is steel with chromium and molybdenum.
43XX—is steel with nickel chromium molybdenum.

The final digits of the code, as mentioned earlier, indicate the approximate carbon content in hundredths of one percent. Lucky for us, it is not necessary to worry much about the finer details of all the codes. We can become familiar with some commonly used steels and get to work.

1020 COLD-ROLLED STEEL SHEET

This is the most commonly used steel in car fabrication. During manufacture, the steel is rolled while it is cold, rather than while still hot. That's where the name comes from. Cold-rolling produces a much stronger steel with tighter dimensional tolerances than the hot-rolling process does.

Cold-rolled steel has a nice clean appearance. It is clear and free from scale or surface distortion. It is strong and can be worked easily, and also is highly receptive to most types of welding. It is available in a number of different *gages*—the industry term denoting thickness.

For example, 1020 cold-rolled steel in 20-gage, which is about 0.035-inch thick, would be my choice to fabricate a car fender. A sheet of 1020 11-gage, about 0.120-inch thick, would be ideal for brackets. The chart on page XXX lists the equivalents of gages and thousands-of-an-inch of thickness.

Because of the exceptional working properties of 1020 cold-rolled steel, it is the best choice of metal for most car construction. It is excellent material for making new floor pans, firewalls, body pieces, gas tanks and brackets. That is by no means all it is good for, but it does indicate its versatility.

AK OR SK STEEL

SK or AK Steel is a steel some people have never heard of, yet it is invaluable to a metal fabricator because it is relatively easy to form. It has excellent *deep draw qualities*—meaning it can be shaped into deep sections easily. This is important on a project which requires the strength and durability of steel, but also requires that it be formed deeply.

The letters **SK** or **AK** indicate the manufacturing process by which the steel was made. During the manufacture, the molten steel was *killed* (indicated by the K) by either the addition of silicon (indicated by the S) or by aluminum (indicated by A). "Killing" is an industry term for stopping the effervescence of molten steel to keep it from combining with oxygen after it is poured into the ingots.

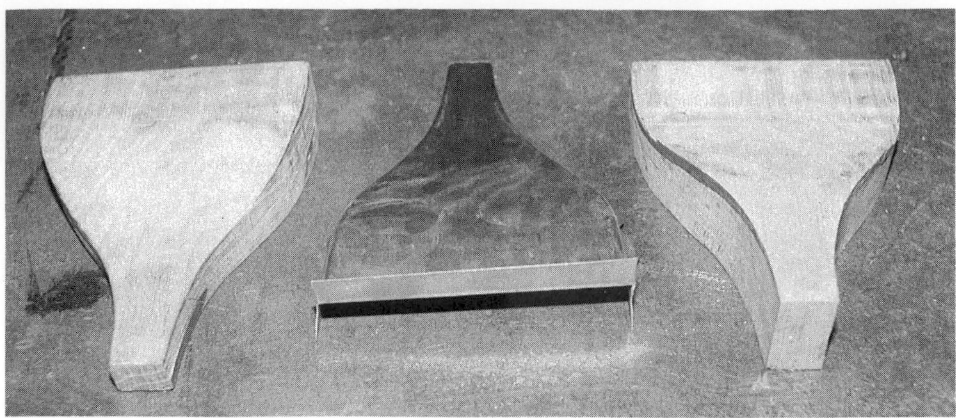

AK steel is easy to hammerform and weld. It's great to use when you're making a part with a lot of shape in it.

Chrome-moly is ideal for use in suspension parts, which need strength. I fabricated all of these suspension components for my roadster from chrome-moly. Photo by Michael Lutfy

The SK or AK steel is somewhat different from other steels made by other processes. The AK or SK steel is of a finer grain. As a general rule, the finer the grain of a steel, the tougher and more ductile the steel will be. AK and SK steels do not work-harden as quickly as other steels. This quality makes this particular steel excellent for hand fabrication.

Hammerformed parts, such as small brackets, are an ideal application for AK or SK steel. The metal responds easily to most shaping and forming techniques, and prototype body panels with many curves are generally made from this steel because of these inherent forming qualities.

CHROME-MOLY

The third steel you need to know is *4130 chrome-moly*. The numbers are the industry code for a steel alloyed with chromium and molybdenum. This alloy has a rated tensile strength of 86,000 p.s.i.—compared to 69,000 p.s.i. for hot-rolled 1020 steel. This translates to nearly a 25-percent difference in strength for 4130.

The major benefit of this steel is that it is incredibly strong. Because of its strength, it's possible to use thinner material to save weight without compromising strength. This is an important feature that makes it the steel of choice for race car or aircraft construction, where the goal is to have as much strength as possible, without adding excess weight.

Chrome-moly has excellent welding qualities, and is available in either hot- or cold-rolled form. Furthermore, both hot- or cold-rolled come in a *normalized* or *annealed* condition. Annealed means the steel has been heat-treated to make it softer and easier to shape. Normalized means the chrome-moly has been *stress-relieved*. Normalized chrome-moly steel is appropriate for a project which doesn't require bending or forming. Annealed chrome-moly is better for something like a lower A-arm, because it can be formed and bent easier than the normalized.

If all this about saving weight without losing strength sounds too good to be true, then you should be aware of some drawbacks. First, and perhaps most importantly, chrome-moly is *very* expensive. It can cost as much as *five times* more than other steels. The price difference can be very significant if the project is a big one.

It can also be difficult to find 4130. Not all steel houses stock it. It is usually a bad idea to special order it from a source that doesn't have it in stock. The special order will usually result in a large price mark-up.

The other disadvantage of working with chrome-moly is that it will dull cutting tools and blades much faster than other steels because it is so strong.

Chrome-moly is used quite often to make highly-stressed parts, like suspension components, roll bars and engine mounts, particularly on race cars. Aircraft pieces, such as engine mounts and landing gear, are often constructed of this material. It is a good, all-around material, but I don't recommend that you do any practicing on it because of the high cost.

CONCLUDING THOUGHTS

Generally, the seven metals I have outlined above are found in the specialized metal shops fabricating race cars or aircraft. These metals are all readily available, though you'll have to look harder for 4130 chrome-moly. By learning the individual qualities of these seven metals, and by experimenting with various projects, you will begin to understand how each one can be used most effectively. Remember, though, to store your metal carefully so it won't be scratched or touched by rust. All metal costs a fair sum of money and it is smart to protect the investment. Before work on any project begins, the metal should be fresh and in perfect condition—otherwise, how can the final results be as good?

This IMSA GTP Corvette uses 4130 chrome-moly for its suspension components. Chrome-moly's high tensile strength (86,000 p.s.i. vs. 69,000 p.s.i. for 1020 steel) makes it the best choice for most racing applications. Photo by Michael Lutfy.

BASIC METAL TERMINOLOGY

In order to communicate with your metal supplier most effectively, and more importantly, in order to choose the material best suited for your particular applications, you need to understand some basic engineering terms used in most metal catalogs.

Stress—Stress is usually expressed in *pounds per square inch* (psi). It represents the load in pounds for every square inch of cross-sectional area. For instance, if a 1-inch square bar is pulled with a 1,000-pound load that tends to stretch the bar, the bar is stressed to 1,000-pound divided by 1 square-inch = 1,000psi. Changing the load changes the stress in direct proportion. Changing the cross-sectional area changes the stress inversely—halving the load halves the stress, but halving the area doubles stress, and vice versa. In simple formula form: S = P divided by A; where S is in psi, P is expressed in pounds, and A is in square-inches.

Although there are different types of stress, I purposely specified a pulling load because it creates a *tensile stress*—a stress that results when a material is being pulled. Why tensile stress? Because stress ratings for metal are tensile stress.

The two types of stress typically listed in metal-stock catalogs are *yield point* and *tensile strength.* Understand that any load applied to any metal will cause it to deform—some more than others. As load increases, so does deformation. When the metal is unloaded, it returns to its original shape—until it is stressed beyond its yield point.

Yield Point—When a metal reaches its yield point, it will continue to deform, or yield, without any corresponding increase in load. Some metals will continue to yield, even though the load may reduce slightly! When this load is removed, the metal will not return to its original shape, but will remain permanently deformed.

Tensile Strength—Often called *ultimate strength,* tensile strength is the maximum stress a metal can withstand before it fails.

Percent Elongation—Percent elongation is the ratio of the deformation of a metal, immediately before it fails, to its original length. For instance, if a 2-inch length of tube is stretched to 2.40-inch before it fails, its elongation is 20-percent (0.40 divided by 2) X 100 = 20-percent. Percent elongation is important because it is an indication of a metal's *toughness.*

Toughness is the ability of a metal to absorb an impact load, which is obviously very important in automotive applications. This is particularly true of many race car components: roll bars, cages and frames, for example. Rather than breaking, the material gives, absorbing energy in the process. As a result, the stressed component will not be as highly loaded or stressed as a less-tough component with a higher tensile strength.

An indicator of toughness that's similar to percent elongation is *percent reduction of area.* When metal fails in tension, it *necks down* or pinches in. Pull a piece of putty apart and you'll see what I mean. The cross-sectional area through the break is smaller than it was originally. Generally, the more a metal necks down—has a higher percentage reduction area—the tougher it is.

Yield Point vs. Tensile Strength—Now that you have an idea of what percent elongation and reduction of area are, you probably won't find it in any metal catalogs. However, another good toughness indicator is the proximity of a metal's yield point to its tensile strength. The farther away they are, the tougher the material.

For example: Mild steel has a yield strength of about 35,000 psi, and its tensile strength is 63,000 psi—1.80 times higher in tensile than yield. At the other end of the spectrum, a popular high-alloy steel has a yield of 60,500 psi and a tensile strength of 95,000 psi—1.57 times higher in tensile. These materials have an elongation of 38-percent vs 26-percent, and a reduction of area of 62-percent vs. 52-percent, respectively. The high-alloy steel is stronger, but the mild steel is tougher.

—Tom Monroe, Registered Professional Engineer, Member SAE.

CHAPTER

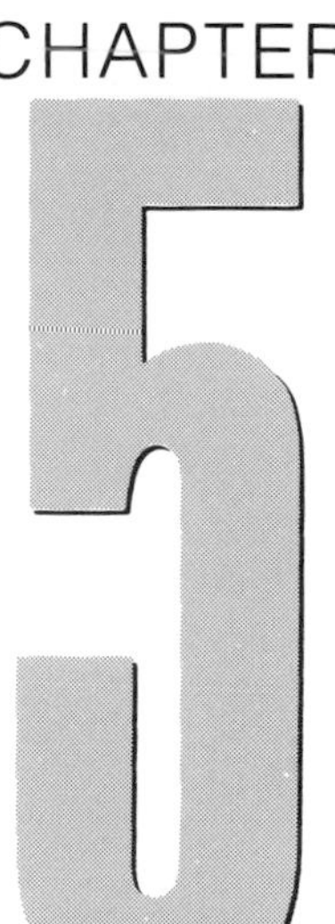

PATTERNS AND LAYOUT

Every metal project begins with an idea that is based upon what you need to make and how to go about doing it. You've already determined where it has to fit and what it has to do. However, the idea is just that—a vague, mental visualization of the end result. To make it clear and concrete, you must first develop a plan for the project before you begin.

Experienced fabricators often think through making the part before they do anything else, visualizing every step to the final result. For a novice, it is easier to plan the project on paper first.

A *pattern* is a plan, a diagram, or a model to be followed in order to make something. By this definition, every plan made for a construction is a pattern. In sheet metal work, a "pattern" is further defined as a detailed paper, chipboard or metal form indicating all the important details of a component that is to be constructed. All preliminary sketches and designs are methods to plan for the actual pattern itself.

PATTERNS

The first step is to gather all of the dimensions of the metal part to be produced. This includes measurements, holes and angles. You'll need to know the width, depth and overall height of the part. Determine where the bend lines are and make sure you have the proper equipment needed to perform them. For instance, if you have a straight bending brake, which can only make bends parallel to one another, but the part needs bends at angles to one another, then you'll need a box finger or combination leaf brake. If you don't have the necessary equipment, then an alternate method of construction will have to be used, such as designing the component with seams and separate pieces instead of bending one continuous piece of metal. For ex-

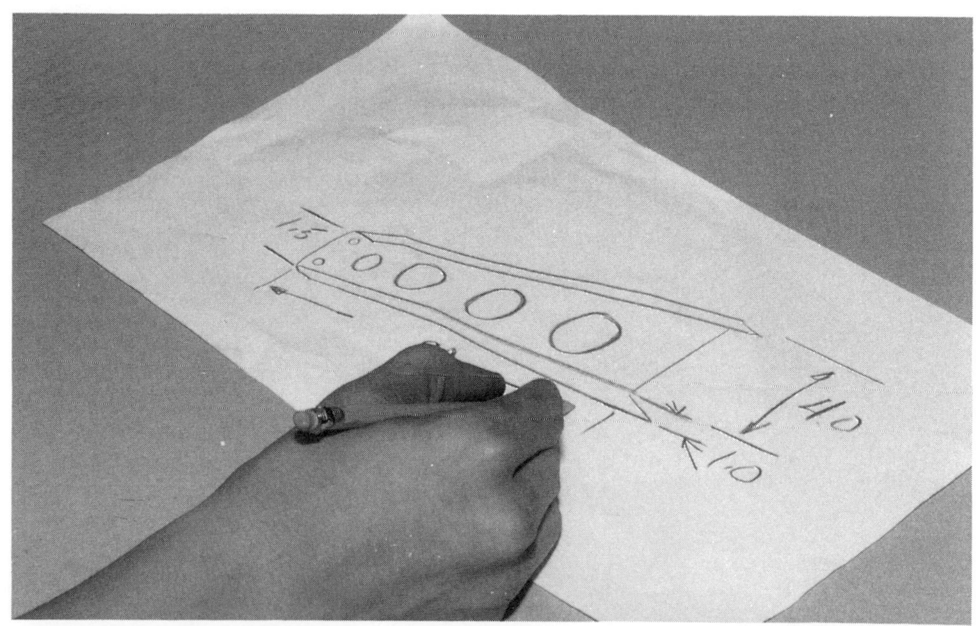

When I'm sketching a pattern I'm also trying to develop a three-dimensional idea of the finished part. A sketch helps me decide if a construction idea is practical. I often make several sketches before deciding on a final design.

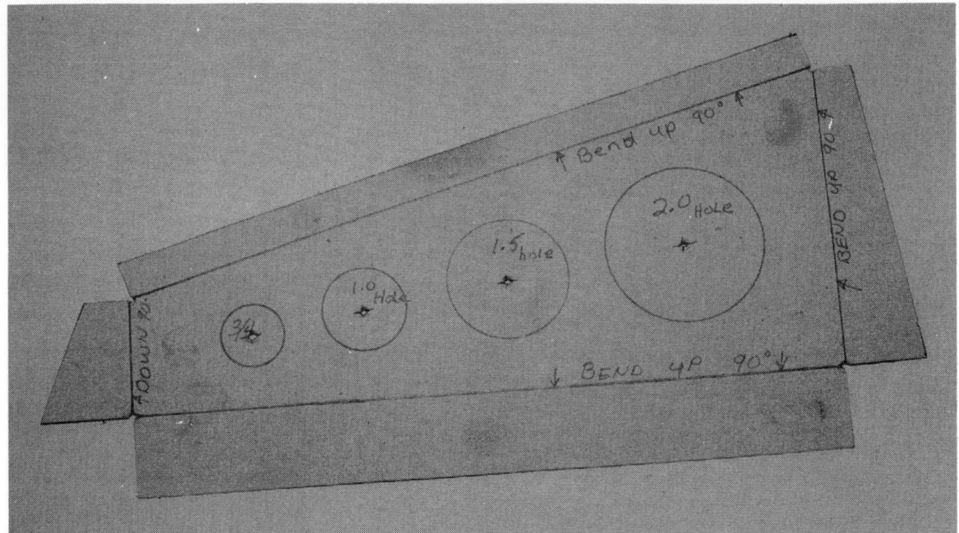

All the information I need to construct a support bracket is clearly written on the chipboard pattern. It shows not only the outline of the part, but also bend lines and locations for lightening holes.

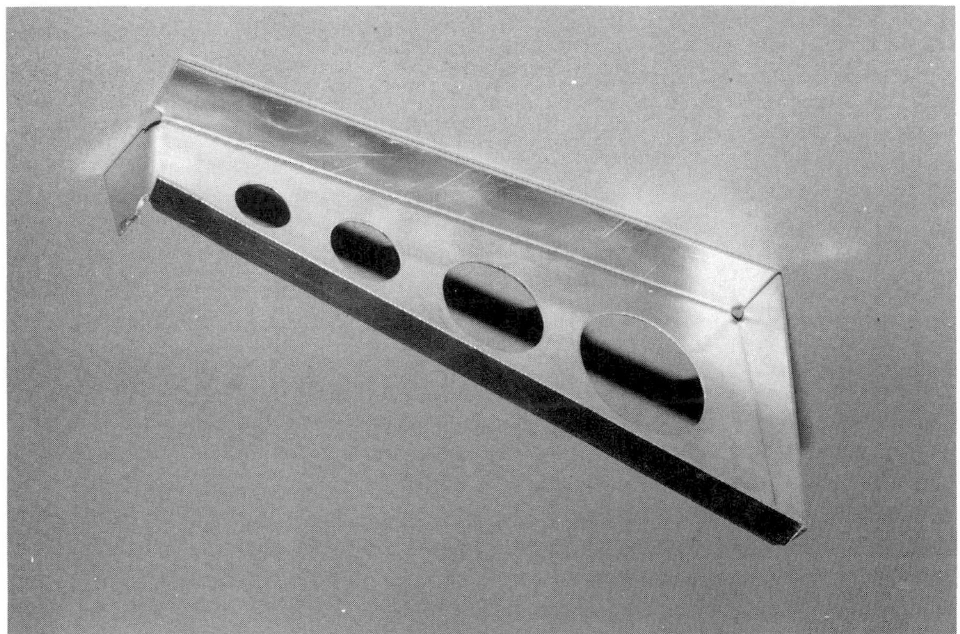
The pattern above produced this support bracket. It was entirely made from one piece of metal.

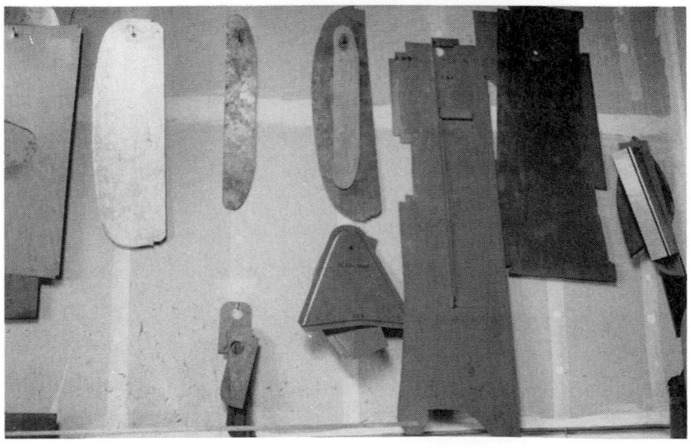
Hanging re-usable metal patterns on a wall keeps them safe, dry and accessible.

ample, it's generally best to make a part by using one continuous piece of metal, folded wherever necessary to make the shape. Part of your planning means thinking ahead on how to make the shape, then deciding how the pattern will accommodate the method in which the piece will be made.

Next, draw the shape of the part on a large note pad, indicating the measurements needed. Make this sketch as clear and detailed as possible. Include all measurements; you'll be sorry if a measurement left out was an important one. Take time to measure, check and recheck. If you have come up with a plan for the part which seems possible for you to construct, recheck the sketch for clarity and accuracy. It is easy to erase a measurement and change it, or to move a line on paper. It is not so easy to correct mistakes if they are discovered when working on the metal itself.

After the sketch on paper, the next step is to determine the type of sheet metal required for the project and order it if necessary. For more information on that, see Chapter 4 on *Types of Sheet Metal*.

Pattern Materials—The paper sketch will then have to be transferred to the appropriate *pattern material*—where the final pattern will be layed out. There are three basic materials used, depending on how often and what the pattern will be used for.

If the pattern will only be used one or two times, then it can be made of *chipboard*—a thin cardboard found at print shops and art supply houses. Poster board will also work, but it is a bit more expensive.

A pattern you intend to use many times should be made of steel or aluminum. A metal pattern will stand up to continuous use for a long time, and will not tear or distort. The more often a pattern is used, the more it will wear and become less accurate. Edges may get irregular. Holes may become elongated or distorted. Though they will last much longer than chipboard, metal patterns should be checked for accuracy once in awhile as well, especially after frequent use.

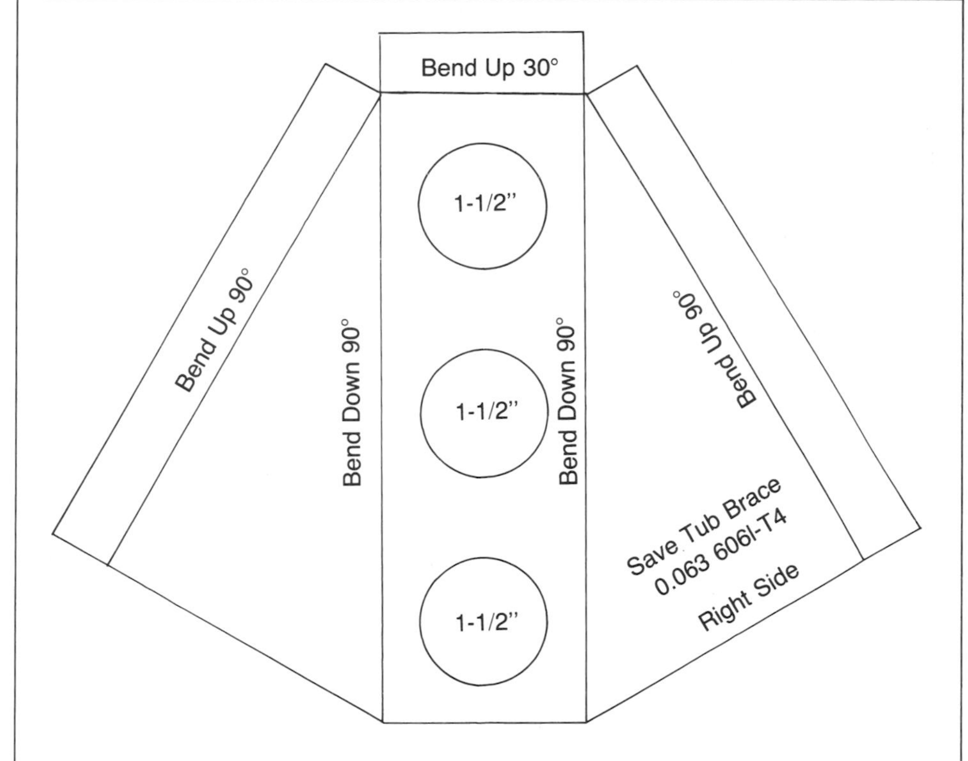

This pattern holds a lot of information, none of it unnecessary. If any of this information were omitted or inaccurate the part could not be successfully completed.

MARKING PATTERNS

Seven important kinds of information must be on any pattern:

1. You must mark the name of the part you intend to make from the pattern. Don't overlook the obvious.
2. You must indicate the type of material you intend to use to make the part from the pattern. Write down the metal alloy and thickness.
3. Mark whether the pattern is for a left or right side, when making parts with a "mirror image."
4. If the pattern is chipboard, be sure to write on it whether you want to save the pattern or scrap it after use. Saving a cardboard pattern can save time if you think you are going to use it a second time.
5. Mark what size holes the transfer punch marks are supposed to be. Generally, transfer punch holes are 1/4-inch or smaller, but they represent actual holes of any given measurement. Mark on the pattern the dimension of the actual hole you need.
6. Mark on the pattern whether the bend lines indicate a bend going up or a bend going down. It is easy to bend the wrong way if the instruction isn't written on the pattern.
7. Mark how many degrees each bend should be. It is not much use to mark a bend up or a bend down unless you also know how far up or how far down the bend needs to be.

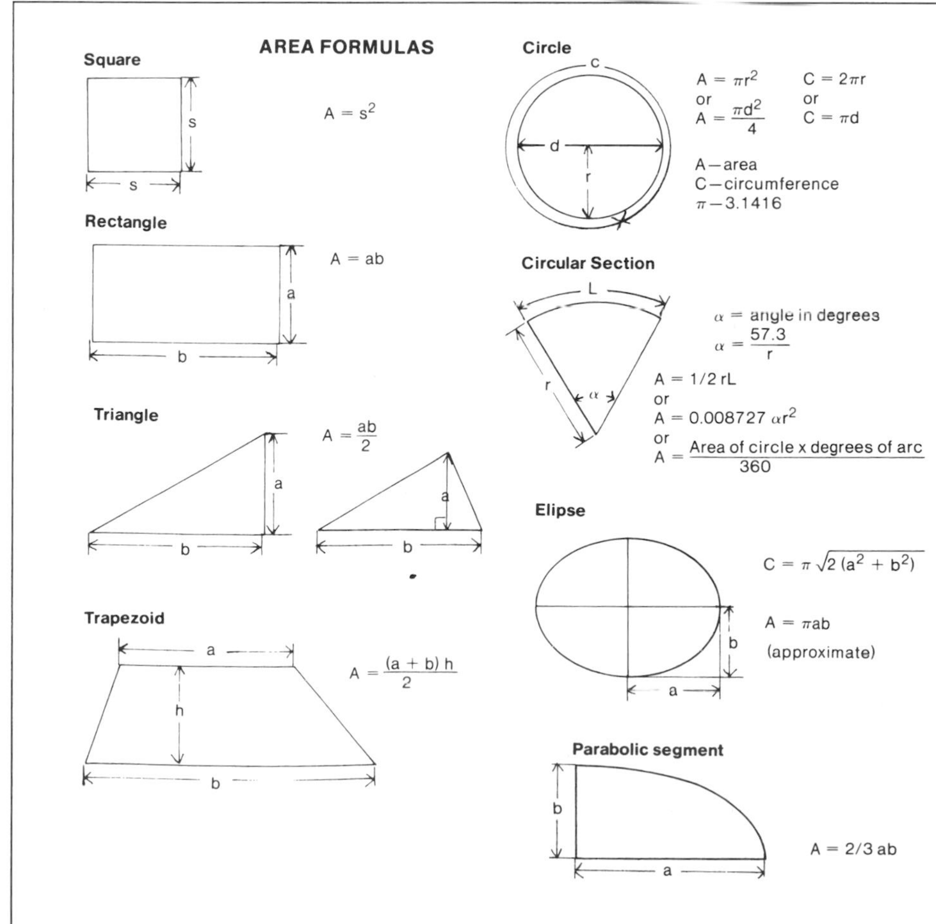

AREA FORMULAS
Square
$A = s^2$
Rectangle
$A = ab$
Triangle
$A = \frac{ab}{2}$
Trapezoid
$A = \frac{(a + b)\,h}{2}$
Circle
$A = \pi r^2$ or $A = \frac{\pi d^2}{4}$
$C = 2\pi r$ or $C = \pi d$
A—area
C—circumference
π—3.1416
Circular Section
α = angle in degrees
$\alpha = \frac{57.3}{r}$
$A = 1/2\,rL$
or
$A = 0.008727\,\alpha r^2$
or
$A = \frac{\text{Area of circle x degrees of arc}}{360}$
Elipse
$C = \pi\sqrt{2(a^2 + b^2)}$
$A = \pi ab$
(approximate)
Parabolic segment
$A = 2/3\,ab$

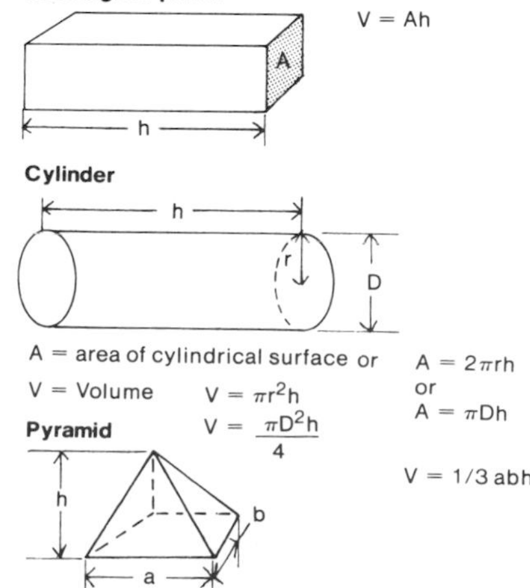

Rectangular prism
$V = Ah$
Cylinder
A = area of cylindrical surface or $A = 2\pi rh$ or $A = \pi Dh$
V = Volume $V = \pi r^2 h$
$V = \frac{\pi D^2 h}{4}$
Pyramid
$V = 1/3\,abh$

Use this chart of common geometric formulas to mark your patterns.

Metal patterns are also frequently used to make a component which has numerous holes. When this is the case, be sure to use at least 1/8-inch thick metal for the pattern to accurately guide a *transfer punch*—a tool with a 1/16-inch-long pointed end used to precisely transfer the center of a hole from a pattern to another piece of metal. The circumference of the transfer punch must be guided by the metal thickness of the metal pattern in order for the transfer mark to be accurate. A thin metal pattern will not guide the transfer punch with accuracy.

MARKING PATTERNS

Every pattern must contain certain essential information. This information serves as a guide during construction of the component, and should indicate every important feature.

It is important that all markings be simple, clear and easy to read. They won't help you if you can't read them. Use black ballpoint pen to make your marks on chipboard. On metal patterns use a scriber or a set of *letter stamps*—metal dies with letters—to stamp markings on the metal pattern. Either kind of pattern needs clear, complete markings. Use arrows to clarify where the measurement applies, or which line you are indicating (see sidebar for Marking Patterns on page 51).

Fitting The Pattern—Once all dimensions have been marked on the chipboard material, the pattern must be cut out and in some cases fitted to the car. Cut out the chipboard piece, or pieces that correspond to the final part. If you have a large piece of chipboard it is possible (and easier) to cut out the pattern in one big piece so it will include nearly all of the bends and flat surfaces. If pieces for each flat surface are cut out separately, simply tape them together along the bend lines with wide masking tape. The result is a taped together version of the final pattern that can be checked on the car to make sure it fits correctly.

If there are problems with the fit, you can make modifications with scissors and masking tape. However, don't hesitate to scrap a pattern and start over. If you do modify it, remeasure the final pattern and make sure it checks out okay.

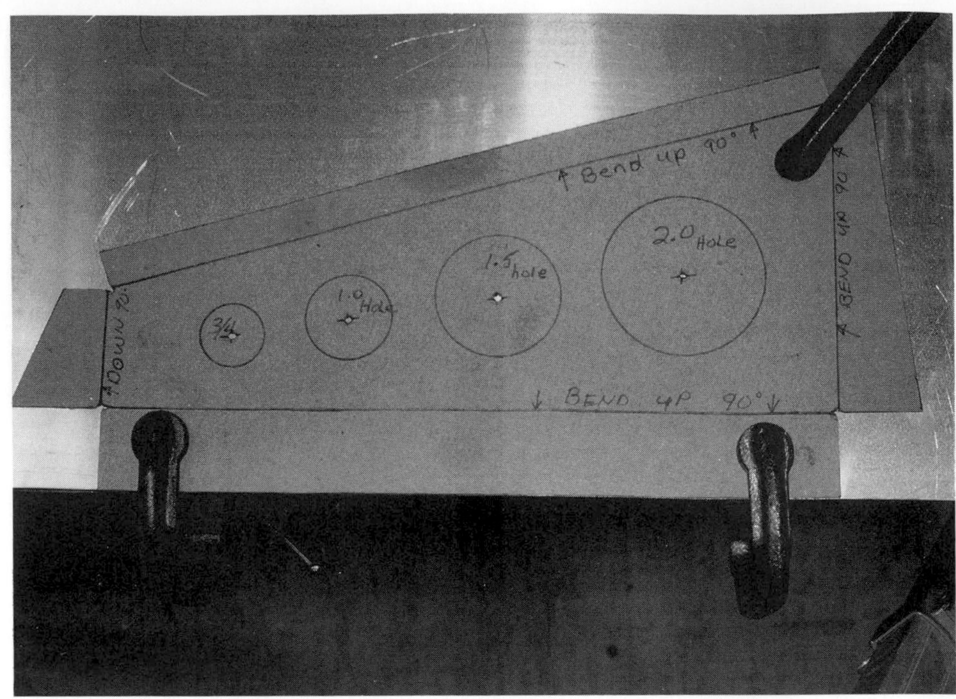

The world's best pattern won't work if it shifts while you're marking the metal. Use plenty of clamps to keep it in place.

Developing a pattern in this manner gives an accurate representation of the finished part. It allows you to determine beforehand where to make the bends or welds. It also enables you to spot-fit problems very quickly and correct them. This process is called *pattern development* because the pattern changes and develops as you cut and fit until it is right. You must fit and refit, measure and remeasure, calculate and recheck calculations. Pattern-making requires care and patience.

LAYOUT

When the pattern is complete, it must be transferred to the actual sheet metal material, or *sheet metal blank*. In order to work with the dimensions, all straight lines, curves, circles and center marks need to be layed out onto the surface of the sheet metal. These lines and marks serve as a guide when making the part.

The layout must be properly located and marked if the final project is to be accurate in size and shape. In order to make all these marks just right, there are some general rules to follow, which will help avoid problems in the end. It is more than a little discouraging to come up to the very last bend on a piece, only to find that one bend line was marked up instead of down.

LAYOUT RULES

The first rule of layout is to have all necessary information *before* beginning. Measurements are the most important, followed by any other details you can think of. I generally measure, recheck and record measurements before I begin marking. Sometimes I make a sketch of the way I think the layout will work, drawing an idea of the layout and noting all measurements and details.

This first rule may be easier to understand if I tell you how it works. For example, if you forget to note how a flange will bend or how large it must be, you may not notice the error until the whole thing has been fabricated. Then when you notice the flange isn't what it should be, you may have to scrap the whole part.

The second major rule for layout is to always work from a *center line*. A center line indicates the dead center of a given part, and serves as a clear base from which to orient all the other lines and dimensions.

For example, if you have a piece of metal which needs two folds 10-inches

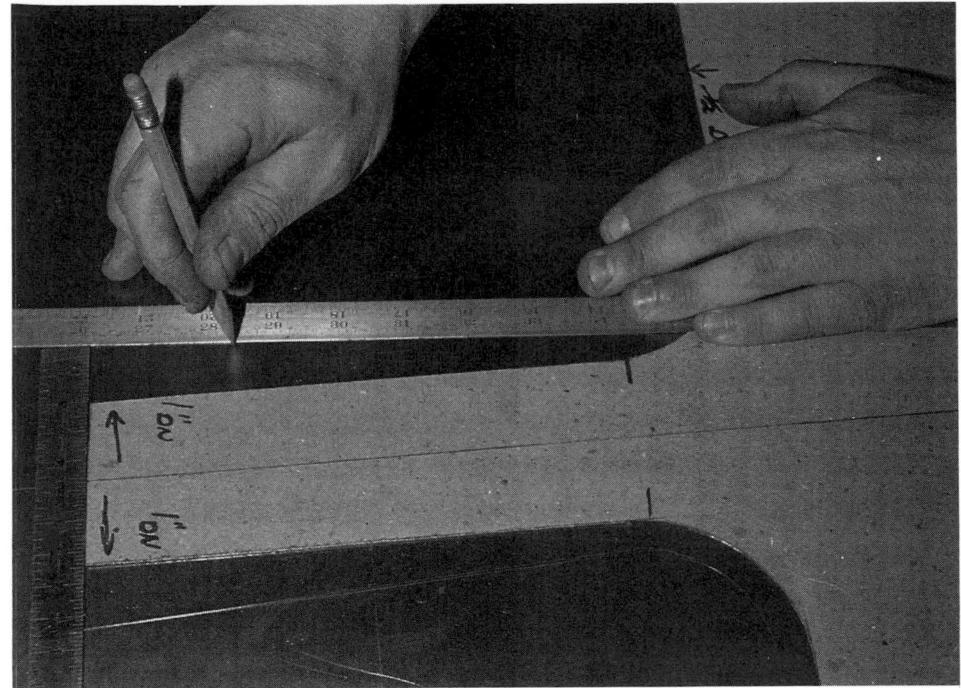

I often use two steel rules to locate a marking point. The lower rule maintains a base line for using the second rule.

Be sure you press toward the rule when using a scriber. It will guide the scriber correctly.

Hold a center punch vertical. Mark exactly where the center should be. Hit it firmly for a good mark.

apart you measure out 5-inches on either side of the center line to locate the fold lines in the correct place.

The third rule for layout is to mark everything *clearly*. Sounds sort of obvious, but if the marks aren't clear it's easier to make a mistake when working along. For instance, if a bend line is marked too faintly it will be difficult to align it in a brake correctly, which could result in an inaccurate bend and trouble.

TRANSFERRING PATTERNS

Bend lines should only be marked in pencil. Do not use a scribe to indicate bend lines, which might weaken the metal and cause it to break later. On light-colored metals such as stainless steel and aluminum, I use a soft lead pencil. For darker metals such as mild steel and chrome-moly I use a white or silver pencil. Keep the pencils sharp, to make a clear precise line on the metal.

Sometimes it's necessary to use *layout dye*—commonly referred to as *Machinist's Blue* or *Dykem*—to coat the metal so layout lines will be more visible. The Dykem is either brushed or sprayed on, and dries quickly. Scribers, scratch awls or dividers are used to mark cutting lines on the dye. The idea is to scratch the marking *just through* the layer of dye, which makes the layout marking very easy to see. You want to avoid scratching deeply into the metal.

Many of the tools used for layout are measuring tools that are fairly simple to use. However, it is critical that you are able to read and understand the graduations on rules and tape measures easily. If you mistake one mark on the rule for another, the measurements will be off, and so will the final result.

The tools most commonly used for layout are steel rules, measuring tapes, squares, dividers, scribes and center punches. Of course, each of those tools comes in a variety of sizes and types. For a more detailed description on these tools and how to use them properly, go back to Chapter 2 on *Basic Hand Tools*.

CHAPTER

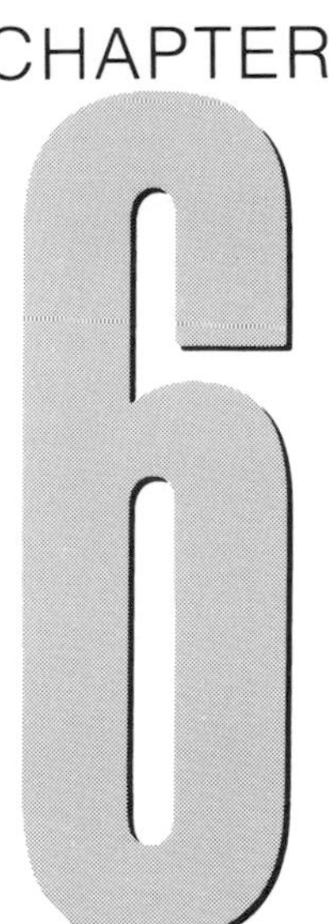

METAL SHAPING

Put simply, metal shaping is taking a flat sheet of metal and transforming it into a curved component. The end result will consist of a shape with compound curves made through a series of shrinking and stretching operations.

Certain techniques are employed with special tools to *shrink* or *stretch* metal to achieve the end result. Shaping is also possible with a variety of large specialty equipment, which is covered a bit later on in this Chapter. In this chapter, I'll be covering some of those techniques and the special hand tools that are used in shaping metal. How the tools work is not difficult. What is hard is learning to use them to get predictably productive results. Practice and hard work are necessary to develop this "feel" for metal shaping, and chances are you are going to get frustrated and discouraged more than once. No one said this was going to be easy.

However metal shaping is one of the most rewarding and gratifying endeavors I know. Did you know that shaping metal by hand is a skill that is thousands of years old? How do you think knights got all that shining armor in medieval times? My family always knows where to find me in an art museum—either in the halls of armor or in the metal sculpture gallery.

GETTING STARTED

It is important to learn to shape metal by hand before you can begin to learn to shape metal with equipment or machinery. This is how the basics are learned, such as how metal reacts when formed with only hand tools, or when it is struck with a hammer. Actually, large equipment and machinery are really an extension of the hand-shaping process.

The tools used in hand-shaping, or forming metal are simple. Steel and wooden hammers, or rawhide and plastic mallets certainly are straightforward enough. Shot bags, slappers, dollies and stakes sound

Anything from an exotic car to a hot rod requires metal shaping for special parts. Mel Swain's custom metal shop in California does many kinds of metal shaping, including the panels for this Ferrari.

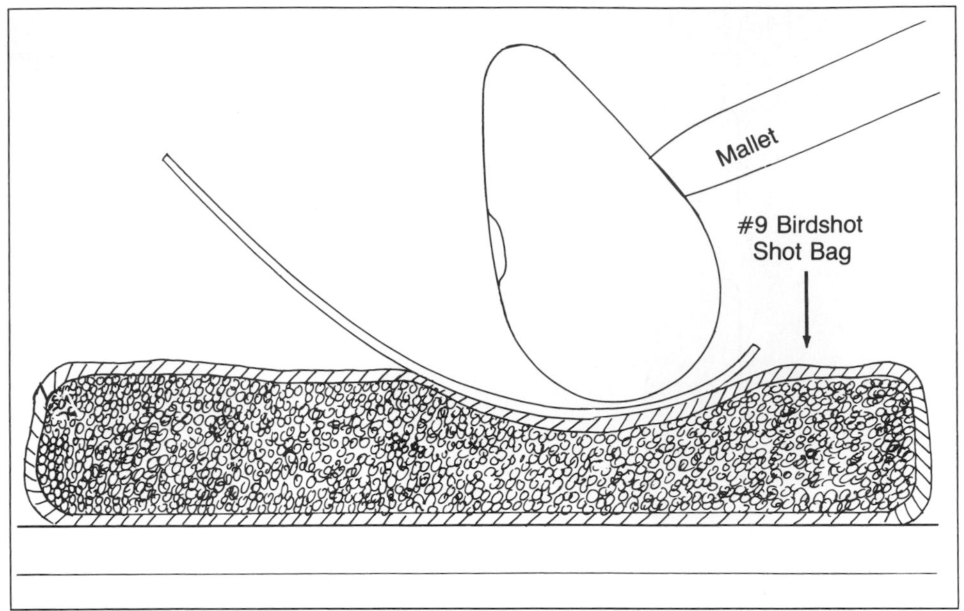

Using a mallet you can form metal over a shot bag. #9 lead birdshot in the leather bag disperses under impact and conforms to the face of the hammer.

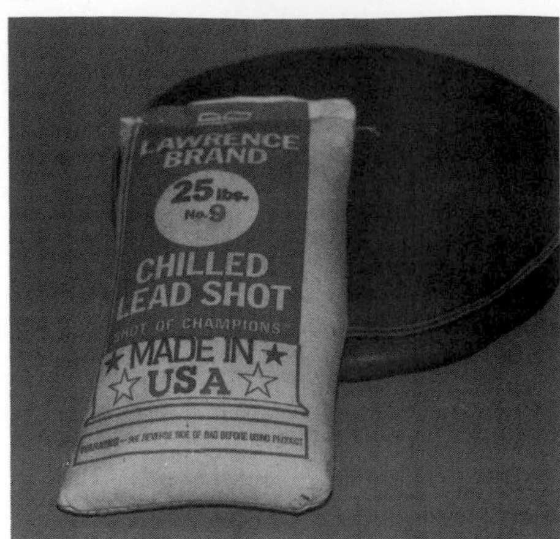

A leather shot bag 12-inch in diameter, 2-inch thick is filled with two 25-lb. bags of #9 birdshot. This shot bag is available through U. S. Industrial Tool and Supply Co.

tricky, when actually they are just simple tools as well. What counts is *how* the tools are used and the skill of the worker. Hand-shaping is not a single action, it's a *process* that takes "shape" over a period of time.

Metal is formed by striking it with a hammer or mallet, directly or indirectly. In order to create the form you have in mind, it is necessary to use force to stretch the metal. The trick is learning *how much* force to apply to get the shape you want. A delicate tap often works better than brute strength.

A good method to stretch metal is to strike it on a surface that will yield to the blow. A *shot bag* is a leather bag filled with lead shot (the same used in shotgun shells, #9 size is best) and sewn shut. The metal is struck on top of it, and the bag gives some support yet yields enough so the metal can form. The shot inside the bag conforms to the hammer's face, and the metal stretches. Shot bags can be purchased from the U.S. Industrial Tool and Supply Co., or you can make one by using high-quality, thick leather filled with #9 birdshot and expertly sewn shut.

Now don't start hand-metal forming by dashing out and hammering on metal over a shot bag. The first steps come before the first hammer blow. Don't set yourself up for immediate failure by taking on a poorly planned or complicated project right off the bat. Try a simple, easy shape first. Believe me, there will be plenty of time for complex projects, ones that will go smoothly if you master the basics first. A simple shape may not be the most useful item in the world, but keep in mind, all that matters right now is the information acquired, and not the product produced. That way, you won't get discouraged. Ironically, the more you focus on learning the skill, the more likely you are to end up with a project you'll like when you're done.

STRETCHING VS. SHRINKING

Forming compound curves in shaping metal requires that you either *shrink* or *stretch* the metal. Sometimes, you do both. *Stretching* metal is achieved by hammering or rolling metal under pressure. The sheet metal thins, elongates and curves. It is critical however, to make sure you don't stretch metal too much, making it too thin. It doesn't rebound easily.

Shrinking metal, on the other hand, is also achieved by hammering, but the difference is that the metal is *gathered* by compressing it or forcing it together. It is much more difficult than stretching. There are two basic methods of shrinking by hand—hot or cold.

Cold-shrinking is done by using the combination of a *hard* surface, such as wood or steel, and a *soft* mallet or hammer. Remember, using a steel hammer over a hard surface will *stretch* the metal, not shrink it. The larger the mallet face, the better. If this sounds confusing, don't worry, you'll get to try it out soon. *Hot-shrinking*, the second hand-method, involves heating the metal first, then hammering. A third method involves the use of shrinking equipment, such as the Eckhold Kraftformer.

BEGINNING PROJECT

Let's make a tear-drop shaped bowl with a flange around the outside. The tear-drop is a good starter project because it involves the most basic process of metal shaping. The metal will be stretched by hitting it quite a bit, but you'll be learning *how hard* to strike it in certain areas to get different results. Anyone can start hammering on a flat piece of metal in the center and get a rough-shaped bowl, but this bowl will have an exact depth, shape and a smooth surface. In other words, you are going to learn *how to control the shape of the metal as you go along*.

Make the tear-drop shape out of 0.050-inch or 0.063-inch 3003-H14 aluminum. Draw a full-scale outline of the tear-drop shape on a piece of paper, both in a top view and side view. The total distance from the top of the bowl to the bottom should not be more than 1-1/2-inches, because this depth is hard enough for a beginner. Next, transfer your drawing to a piece of chipboard. It should look like the pattern shown in the illustration on this page with the same dimensions but drawn to full-size. Cut out the pattern and set aside.

TEMPLATES

A *template* is used to help check the accuracy of the part as you progress. It is generally made of light chipboard and represents the final shape you are working toward. For the tear-drop shaped bowl, you'll need templates for both directions—from side-to-side and front-to-back. A main center line template and two or three templates at 90-degrees to center line should be enough to check your work.

To make a side-to-side template for the bowl, start with a piece of chipboard three inches wide and eight inches long. Using the illustration as a guide, draw the center line from the midpoint down to one of the long sides, mark up and on the center line 1-1/2-inches, then out on either side of the center line 2-3/4-inches. Connect all three points by drawing a smooth arc between them. Cut out the arc, and you have the template for side-to-side checking.

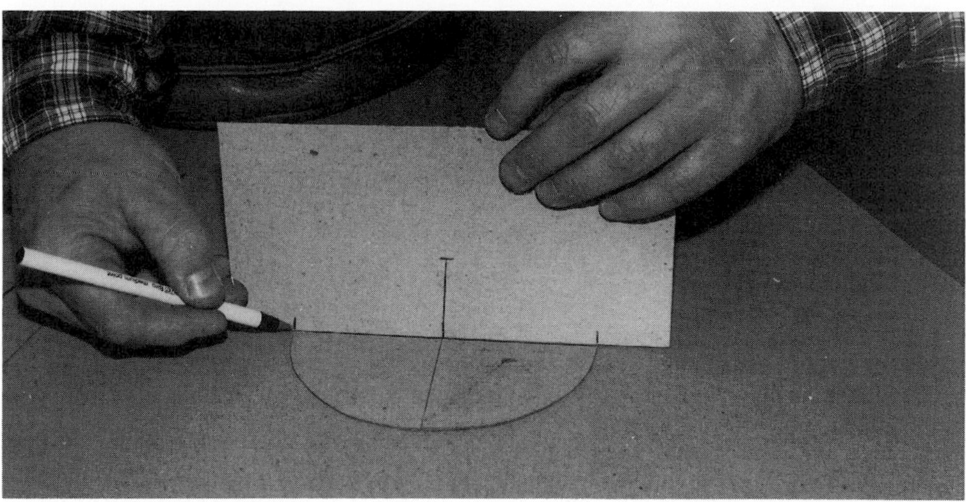

I've already drawn the base shape and I'm working on the mid-section shape. I'm keeping it 1-1/2-in. deep so it will be easy to form.

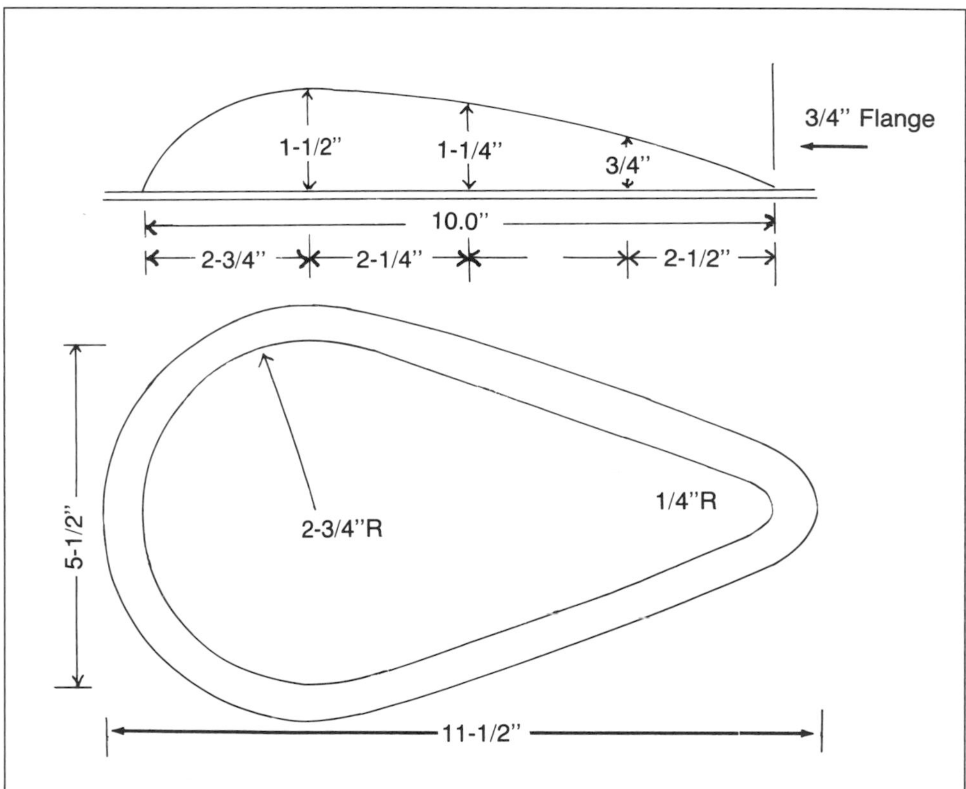

This is the pattern to be used for making the tear-drop shape bowl. I've included a top and side view. The measurements are to be drawn to full-scale.

ANNEALING

You can make the aluminum easier to shape by *annealing* it. With aluminum, annealing is done by covering the area with carbon from an oxyacetylene torch set with excess acetylene. Then the flame is adjusted to neutral and the area is heated evenly until all of the carbon is burned off. Let the aluminum cool at least 200-degrees by standing in the air. Depending on the room temperature, this 200-degree cooling may take five to ten minutes. It is fairly quick. After cooling this much, you can speed the final cooling by quenching the metal in cool water, or, if the panel is too big, a sponge and a bucket of cold water will do.

I suggest you anneal the aluminum for this project so you will have an easier time. I wouldn't want you to give up on shaping because it seems too hard. Annealing makes the metal softer and easier to form.

MARKING THE METAL

Take your pattern to the cooled, annealed aluminum sheet. Mark the pattern onto the metal. Use a #2 or softer lead pencil so you don't scratch the metal surface. Trace outside the pattern. Then mark exactly 3/4-inch outside the pattern outline, all around the shape, and draw *another* tear-drop shape. This will allow for a 3/4-inch flange around the tear-drop. One reason for the flange could be to mount the metal part, using rivets, to the bottom of a jet plane wing, or to the hood side of a '32 Ford roadster needing clearance for hemi heads. Another real

I began hammering at the center and worked evenly toward the outer edge inside the flange. Don't hammer on the flange yet.

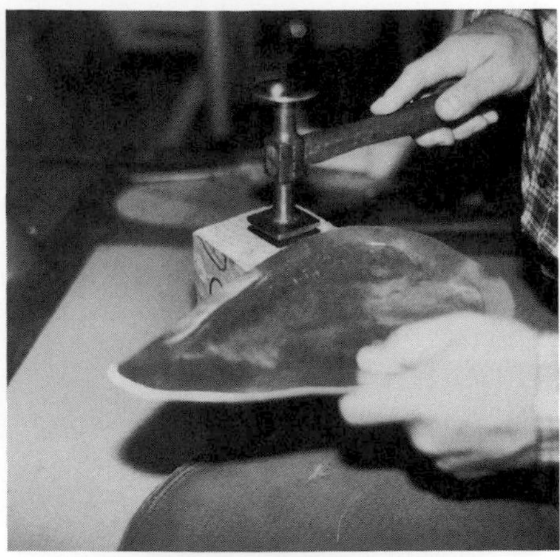
The flange will get wavy, so you'll have to shrink it by holding it firmly over a wooden block and hammering out any waves to shrink the excess metal.

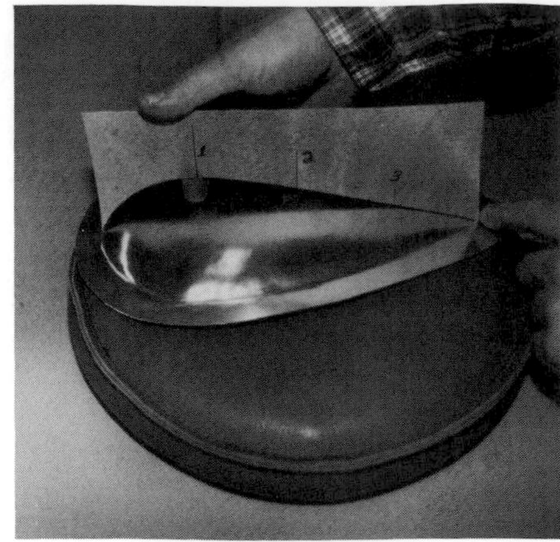
A quick check with the template shows more stretching is needed at the round end of the teardrop.

good reason for the flange is that it will help control the part as you form it. You may decide to trim off the flange when you are done. Trim it off when the tear-drop is going to be welded to another part like a hood.

Cut out the outer shape in the band saw or with hand shears. File off all the edges very smoothly so you won't scratch your fingers or damage the shot bag as you work. Take time to make all of the edges smooth.

HAMMERING

Okay, now we are ready to make a beautiful tear-drop shape with a 3/4-inch flange around it. Hold the aluminum directly on the shot bag. Choose a large, curved-face steel hammer, or a mallet with a large curved face without any mars on it. *You must not damage the aluminum surface in any way.* Damaging the aluminum surface will mean unnecessary extra work smoothing it later.

Hit with a strong, firm blow at the center of the tear-drop, and work up to the *edge* of the flange. *Don't hit the flange*. Leave the flange alone for now. Hammer as close to the flange of the tear-drop as you can. Move around the outside of the shape until you come back to where you began. Now move the hammer up a distance equal to 3/4 of the diameter of the face. This will allow you to overlap the first area of the hit. Keep the blows even and of the same strength. Overlapping, equal

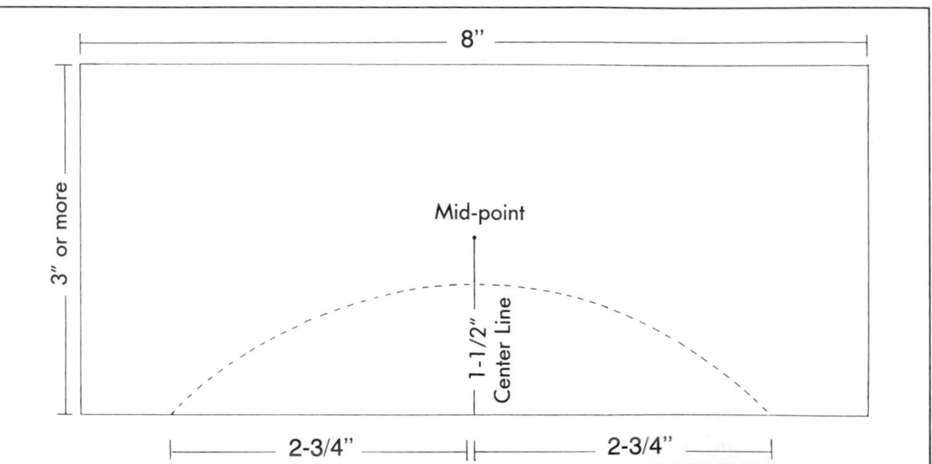

This drawing should help you make a template for the bowl.

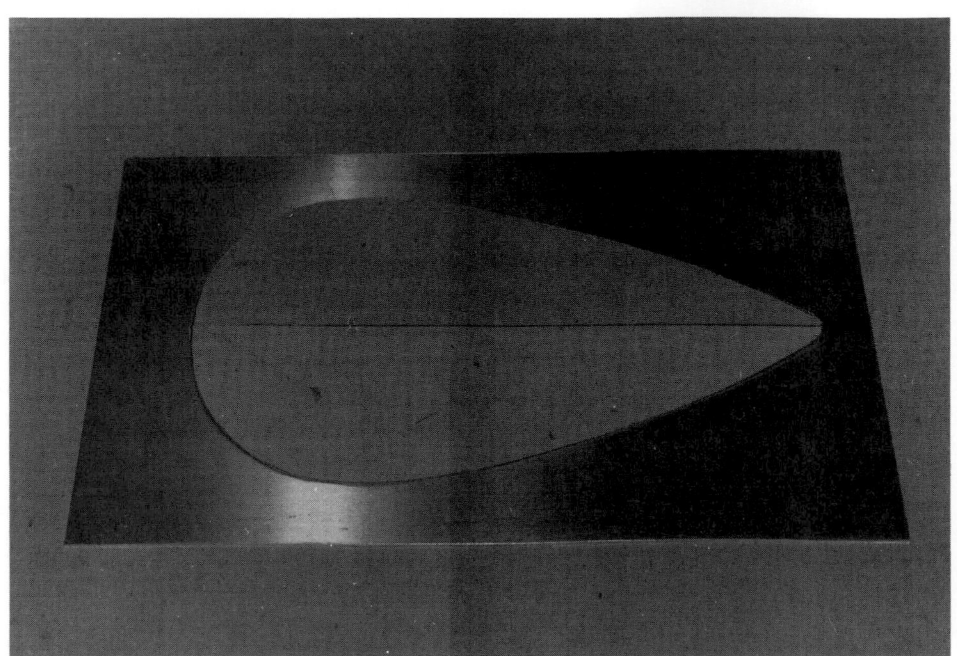
Lay the pattern on the metal blank. Mark carefully, allow an extra 3/4-in. for the flange and cut it out with a band saw or hand shears. File and sand smooth the edges before proceeding.

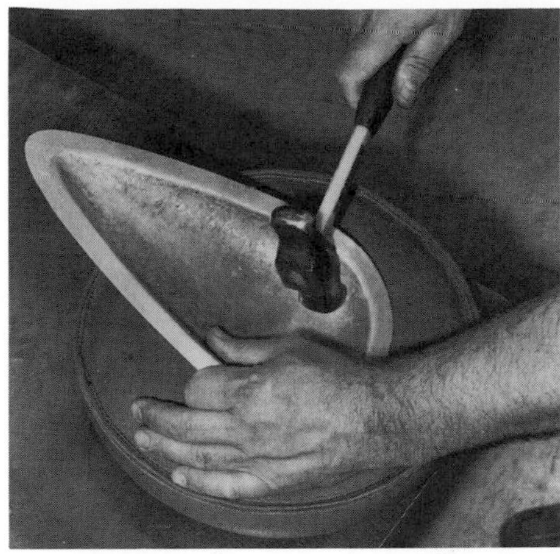

I continued stretching using overlapping, hard hammer blows. Photo by Michael Lutfy.

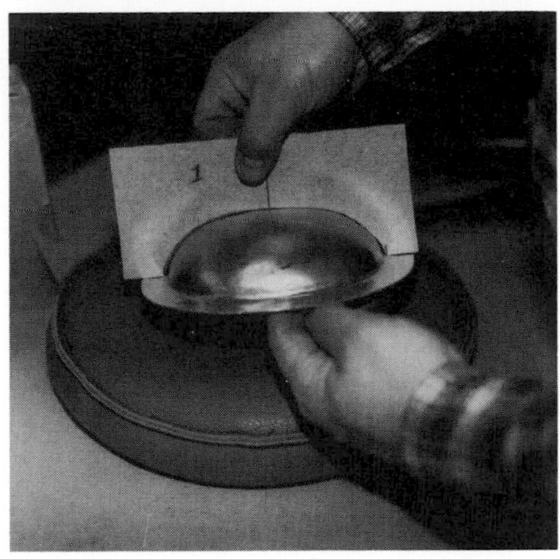

A final check with the templates showed the part conformed to the desired shape.

I used a small hammer to planish the part over a dolly held in a vise. I also used a large, stainless steel dome underneath the round end. Another alternative is to use a small steel slapper. Its shape means the blow to the metal covers a longer area. That makes it easier to overlap blows when planishing. When all the planishing was done, the part was very smooth. The whole reason it was so smooth was the small, overlapping blows I used with the slapper.

blows will make the surface smoother, without lumps or bumps. After you've completed a "lap" around the tear-drop, *turn it over and check it with the template*. Do this often to make sure that the shape is progressing according to plan. Continue overlapping until you get to the flange line. The larger end will need to be hit harder, as this section is deeper and needs to be stretched the most.

The hitting and stretching must be uniform. The templates will determine where depth needs to added, or keep you from stretching too much.

As you form the part, the flange is going to become wavy. but it must be kept straight. To shrink it, hold the flange firmly on a flat wooden surface and hit it with a large, flat-faced wooden mallet or hammer. Hit the high spots to force them to gather. Firm blows will force the metal down flat. Remember to concentrate your hitting on high spots. The low spots will take care of themselves if you handle the highs. The large, flat face on the mallet will give maximum shrinking. This type of hammering will shrink excess metal by working out the waves and flattening the flange. Shrinking and stretching, the two most basic skills of hand metal forming, are usually used together in this way to shape metal components.

Forming the rest of the tear-drop shape is just a repeat of this process: stretch with overlapping firm, equal blows; check with template; shrink flange with wooden mallet; check with template; and so on. Keep repeating these steps until the form matches the templates. If you skip one step, then it will be that much more difficult to final-finish, or *planish* the shape.

PLANISHING

It is unusual for a part to be very smooth on the first try. Quite often the surface will have to be *planished* after the part is formed. "Planish" literally means to smooth by hitting. Though this sounds paradoxical, the surface will be evened out and made smooth by hitting it with a hammer. Planishing is the second stage of forming the part after achieving the shape you wanted.

Planishing works well if two basic rules are followed:

1. First, hit the work many times and cover every square inch of metal. However, don't hit it hard. Hitting it hard will cause stretching.

2. Second, a steel hammer or a steel slapper must be used with a steel dolly or dome, which must be very close to the shape of the bowl.

Hand-planishing takes time to learn. Take the time. Use many soft, overlapping blows with a small, flat-faced hammer. Watch to see how the metal responds to repeated soft blows.

FINISHING

Finishing the surface is the final step in metal shaping by hand. This is achieved with files and a disc sander with a soft pad.

Go over the entire surface with a file. I recommend a 10-inch flat/half-round, second-cut file. The file will show any high or low spots still in the part, but if you've really taken the time to planish carefully, then there shouldn't be too many. If you do find some, work them out by re-planishing the problem areas. Remember to use soft, overlapping blows.

When it is all smooth you are ready to use a soft-pad sander. The soft-pad sander is a great tool for finish work. The pad is 10-inches in diameter, made of foam rubber where sandpaper is glued. The advantage to the foam rubber is that it allows the pad to give as it is used. It won't gouge or groove the work easily. The foam pad allows the sander to conform to the curve of the metal surface. It's a great tool, I've used one for many years.

When you begin to sand, start out with an 80- or 100-grit paper. Then move on to finer grit paper. The finer the grit you use, the smoother the met-

I clamped the part firmly so I could file it. Long, even, overlapping strokes from the top made the filing uniform. Filing the round end required a different technique. I used long, curving strokes following the contour of the part's curve.

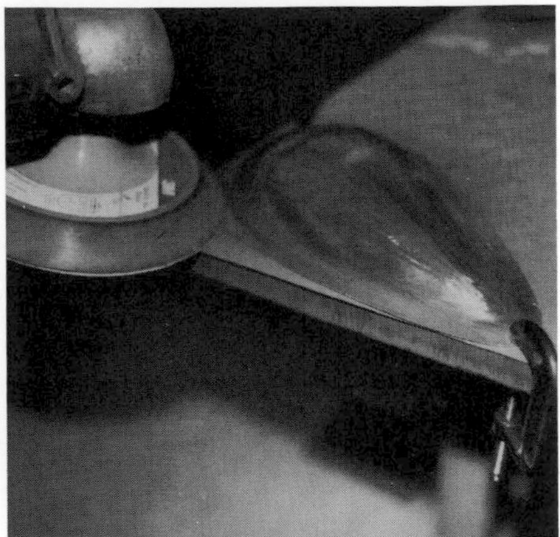
The soft pad sander can follow the curve of the part and also sand the flat flange with equal ease.

The top surface of the finished part shows a flawless, mirror-like finish. I started with 100-grit paper and worked my way up to 600-grit paper to get this finish.

al surface will be.

It is a good idea to use a little wax on the sandpaper. This will keep the aluminum from *loading* in the sandpaper. The soft aluminum sanding dust can clog the sandpaper easily.

After the soft-pad sander you can sand the metal further by hand for a really beautiful finish. The more time and energy spent polishing the surface, the better the finished part will seem.

There are all kinds of projects that can be made by hand-forming. Just remember to keep the project comfortably within your skill level. As you progress, you can make other more complicated and larger parts. Air scoops and body parts are a few of the more ambitious projects possible.

INTERMEDIATE PROJECT

The shaping of the tear-drop shaped aluminum bowl was done entirely with hand tools and by stretching the metal only. Now let's do a project which requires using more sophisticated tools and techniques—such as a hand-formed aluminum air scoop.

We're going to use a sheet metal shrinker, a combination rotary machine (more commonly known as a beader) and a combination sheet metal brake. These machines increase your ability to hand-form more complex metal projects.

Keep in mind this is a learning exercise. Don't take it too seriously if it doesn't come out just right. Start over and try again. Making mistakes and correcting them is part of the learning process. Don't get discouraged.

In addition to the equipment already mentioned, you'll need a T-dolly and hammer to round off the scoop opening by hand-forming, and a soft-pad sander to do the finishing.

THE PATTERN

When I first thought of using this scoop as an exercise in the book, I wanted it to be simple and attractive. All I had was an idea, so I started by making simple sketches with pencil and paper to visualize the shape I wanted. I made several before I knew what I wanted the finished scoop to look like. The next challenge was to make a pattern to produce a scoop resembling the sketch.

Making a pattern follows a sequence. The sketch is used to define the general shape. Next, you'll have to decide how big it will be and establish dimensions of height, width and length. I wanted a medium-sized scoop, so I settled on an 8-inch width and 2-1/2-inch height as the size for the front opening. The length would be about 12-inches, with a taper on the top view. I knew these dimensions would produce a scoop like my sketch.

The next step is to determine the *degree of the curve* from the front to the rear of the scoop. To do that, I made a few curved lines on cardboard that started out flat and curved down, and were each approximately 12-inches long. I chose the curved line that resembled the one on the sketch the most and used it as a template to curve the scoop.

I put all these ideas and dimensions together to make a pattern that included the top and sides of the scoop only. The sides include an allowance to form a mounting flange after the metal shaping is done. The mounting flange will be bent in the final step of forming the part. By including the flange in the overall pattern I avoid having to add a flange made from a separate piece later on. It saves work and makes a seamless part. I also included 1/2-inch of extra metal along the opening edge. This allowance will be rolled under the edge for a smooth opening later on.

By this point, you should begin to understand why it's important to think the whole job through before you begin to cut or shape metal. By planning carefully and thoroughly, you'll increase your chances of producing efficient and satisfying parts rather than expensive scrap. Use your head before your hands or tools.

The pattern is now complete. After drawing in a center line and the important bend lines, the pattern is ready to be layed out on the sheet metal blank.

Layout—Lay the pattern on a fresh, scratch-free piece of aluminum. For

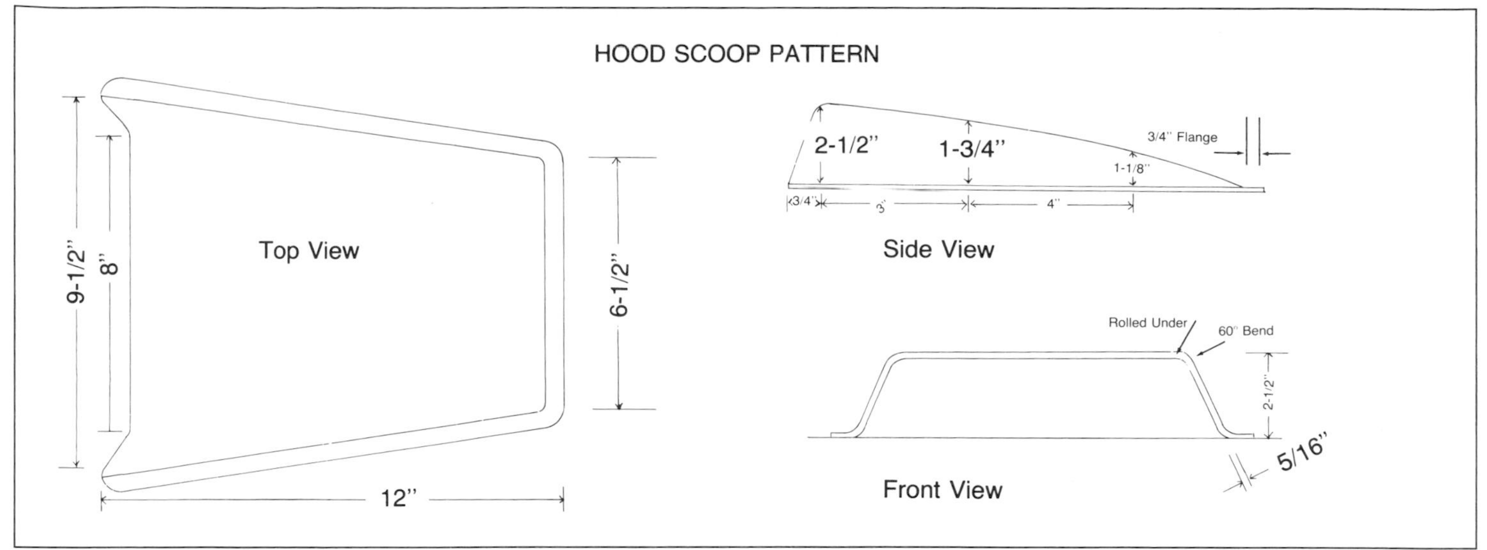

Use this pattern to make the hood scoop in this exercise. Once you get a feel for the process, you may want to make your own using these measurements as a guideline.

this project, I recommend choosing either 0.050-inch or 0.063-inch thick 3003-H14 aluminum sheet. This material is best for this particular project.

Mark all the pattern lines with a soft pencil. Cut out the metal on a band saw. File off all the burrs from the cutting, so you won't cut yourself handling the metal. If you skip this step, you may be reaching for the Band-aids.

FORMING

The first step to forming or shaping the scoop is to bend the sides. A combination sheet metal brake is best for this step because it allows for the use of a 1/4-inch radius die. A small radius die produces a rounded rather than a sharp angle, which I think will look better in this instance.

I bent the sides of the scoop at 60-degrees over a 1/4-inch radius die in the sheet metal combination brake. I decided 60-degrees would approximate the degree of bend in my sketch and the 1/4-inch radius die would produce a soft curve at the top of the scoop.

Let's clarify something. I used the radius bends because I really wanted to have a soft curved angle at the top of the sides. It would have made little functional difference in the scoop if I'd made sharp bends on a brake. I made rounded bends to achieve the look I wanted. This is an aesthetic choice you'll have to make for yourself.

The chipboard laying on the aluminum blank is a simple, two bend pattern. The next step is to mark the pattern on the metal and cut it out.

To get my 60-degree bend with a 1/4-inch radius, I chose a die 1/2-inch in diameter and 15-inches long, mounted in the combination brake. The bending went smoothly and quickly. I clamped the work down and bent the sides up. I made sure to stop at 60-degrees on both sides by checking the angles with a protractor to make sure they were equal and accurate.

After bending the sides, the part still didn't look like a scoop. The sides looked fine, but the top needed help. It was flat, not curved. It had no real shape to it.

In order to shape the top, I had to shrink the sides. Before starting to shrink, I found the template I made for the top and put it right next to the shrinker so it would be handy. One quick look at the template and I could see most of the shrinking had to be done at the back of the scoop. The opening end would need no shrinking.

Be careful bending the blank. Don't go past the degree of bend you want. I decided 60 degrees would work well.

The chipboard template is exactly the shape I want in the curved top of the air scoop. I compared template to the blank as I bent, to see if I had the correct shape.

While shrinking the sides, I often stopped and let the shrinker just hold the metal as I checked the curve with a template.

The curve started about 4-inches from the opening. It started to gradually drop until about the half-way point. It fell more steeply from there.

I started to shrink at the opening end and moved toward the tapered, closed end. As one side would curve a little I turned the part around and did the opposite side. This made sure the sides would stay even as I shrunk them. As I progressed, I checked the curve often, using the template. Most of the shrinking was done in the rear 6-inches of the scoop. The template told me when all of the shrinking was complete. The important thing was to shrink both sides equally.

Make *sure* both sides are equal. Any metal in excess of the 3/4-inch flange allowance must be trimmed, using either aviation snips or a Kett electric shear. Be sure to deburr and clean all trimmed edges. Use a file to remove all the flaws and uneven spots along the whole edge.

The shrinker leaves marks in the metal, and the next step is to quickly remove all of these marks. I did this with a soft-pad sander, and 80- and 100-grit sandpaper with plenty of wax. The soft-pad sander removes the marks quickly.

Remember the sides are long enough to allow for a mounting flange which will soon be bent. It is best to sand and smooth the sides *before* bending the flange. Obviously, it is easier to metal finish a flat surface than one with an angle bent into it. When the sides are all smooth it is time to mark where the flange is to be bent.

After the shrinking, I clamped the scoop to the bench so I could cut off excess metal with Aviation snips. Next sand all burrs and file the edges.

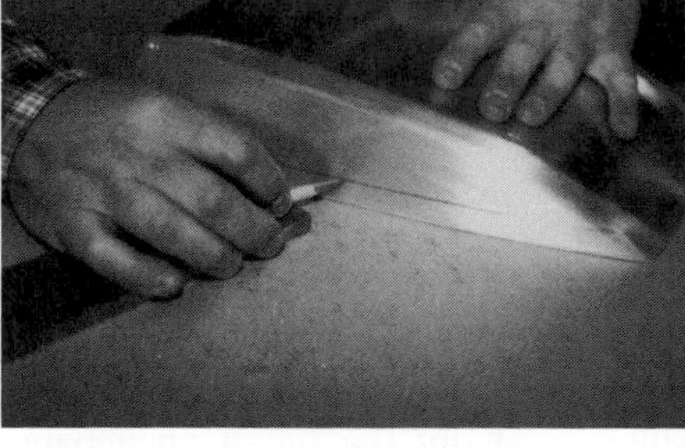

Sometimes I mark where metal is to be trimmed off by using a height gage and scribe. But you can't use a scribe to mark a bend line, or you'll risk metal fracture. I marked the bend line for the flange by stacking metal bars 3/4-in. high. They supported my pencil as I swung it around to mark the 3/4-in. line for the flange.

Flanges—The scoop is meant to be attached to a flat surface, which means a straight bending line will have to be marked on the sides of the scoop to form the flange.

Start at the open end of the scoop and measure from the top down. Mark a point on the side 2-1/2-inches down from the top. There should be at least 3/4-inch of metal below your mark. Take a ruler or straight edge and hold one end of it on the point you marked. Hold that end firmly in place and move the other end of the ruler up or down until you find a place with 3/4-inch of metal below it. Use a soft lead pencil and mark across these points on both sides. When both sides are marked, connect the two lines across the back edge. This is illustrated in the accompanying photo sequence.

There should now be a smooth pencil line marking where the flange can be bent. There should also be a smooth bottom edge. The mounting flange can now be bent up, but because this bending is along a line going around the part, it is best to use a beader with special dies to make the bend.

Bending With Beaders—This type of bending requires a helper. One person works the crank of the beader to turn the rollers and the other person handles and guides the metal through the beader.

If you have never worked together before, take time to practice on scrap aluminum. You must work as a team. The person who cranks the beader can ruin the part by cranking too fast. The guiding person can ruin the part by letting the die run off the line. The faster one person cranks, the harder it is to keep the beader on the marked line. This is especially true when bending curves or turns.

Take some time to get your act together. Practice and see how to fol-

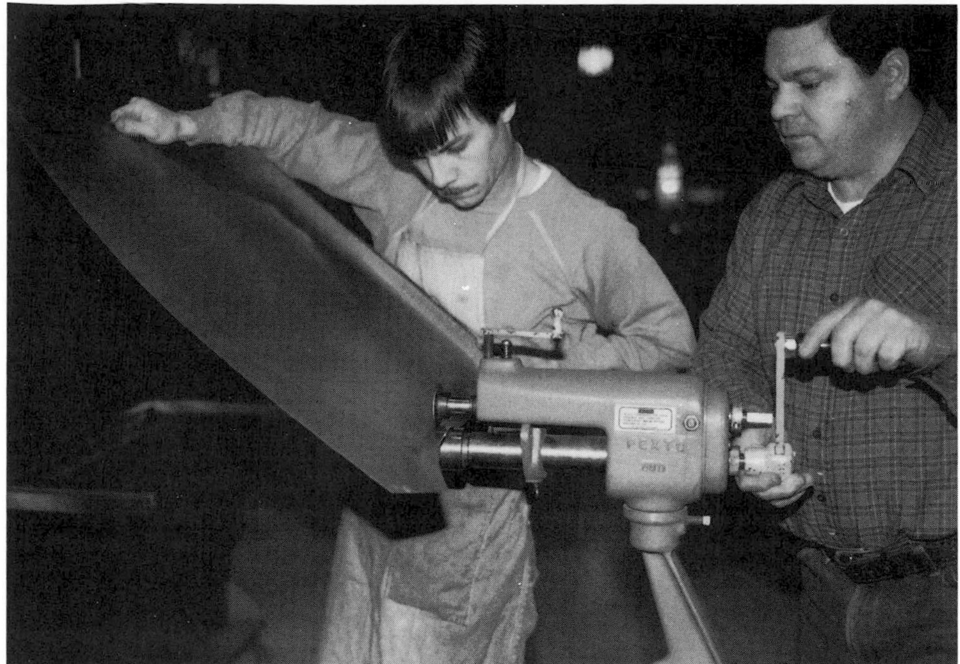

Steve Miller is actually indenting the metal so he can bend a curved flange. I'm just cranking to provide power while he tells me how slow or fast to crank. This teamwork can produce great results.

After bending the flange on the beader, I corrected any surface irregularities by tapping with a small hammer over a flat wooden surface.

I've cut back the first 1/2-in. of flange so when I roll the edge the flange won't interfere.

low the line together. Operating the beader takes cooperation, communication and practice. It gets easier the more you work together.

Bending with a beader takes a special set of dies. The top die looks like a round blade or disk about 2-inches in diameter. The edge touching the metal is pointed, but it isn't pointed enough to cut the metal. The bottom die is a round barrel with flat sides, about 2-inches in diameter and 1-1/2-inches long.

Start bending at the scoop opening. Go all the way around. Bend the flange in a few stages. Bend a little more each time. Go completely around the part without stopping. Then re-insert the part and go around again. This process makes a smoother bend. The only reason to stop bending is if you have accidentally gone off the line.

Remember this is a team job. Both the cranking person and the guiding person have to keep their eyes and attention constantly on the metal bend line and watch what it is doing. The person who is cranking must watch closely so he can stop immediately if his partner has wandered off the

It takes teamwork and steady eyes to bend a straight flange with a beader.

marked line. The guiding person has to tell the cranking person instantly when he wants to slow down, or if he needs to stop and correct.

The flange must be bent evenly all the way around, gradually and in stages. Don't let any one area of the flange bend more acutely than another. You know the flange is bent all the way when you check it with a protractor and it all measures equally 60-degrees. Check it on a straight surface to make sure it is flat. You can straighten out problem areas with a flat hammer. Be sure to tap the flange softly so you won't stretch it.

Now that the flange is bent and you're sure it is straight, it is time to finish it off by rounding the edge at the opening. This final step is really worth the effort. Any piece of metal work looks better without raw edges.

Rounding The Edge—To round off the edge, you need a small T-dolly—about 3/8-inch in diameter by 1-inch long. You'll also need a soft-faced hammer. The objective is to roll the edge over and under.

Start with rolling the edge over. Hold the metal over the T-dolly. Start to tap the metal down, but move all along the full length as you tap. Move the metal evenly as you work. When the entire edge is bent down at least 90-degrees, the T-dolly will be used in a different manner.

Now you want to move the edge under. Hold the scoop upside down against the bench or some other firm smooth surface. Hit the metal back even farther. Hold the T-dolly in against the bend at a 45-degree angle. The T-dolly supports the hammering. Keep the hammer blows at the edge of the metal to keep the rounded edge thick and full-looking. Be careful not to squash the edge by hitting it too hard. When the edge is 180-degrees, stop and check to make sure it is smooth. You may want to planish the surface with a few hammer taps or a few strokes with a file before proceeding to the next step.

Finally, sand out any scratches. This can be done by hand with 100-grit sandpaper to give the scoop a professional appearance.

I encourage you to design and fabricate a different scoop, one of your own dimensions and designs, once you understand how this process works. Whether or not you hand-form a number of scoops isn't important. What you should have learned by this point is how the metal behaves and how to control it with hand tools.

I started rolling the edge with a T-dolly held in a vise. I worked evenly all across the top and sides hammering gently, until the metal was halfway rolled under.

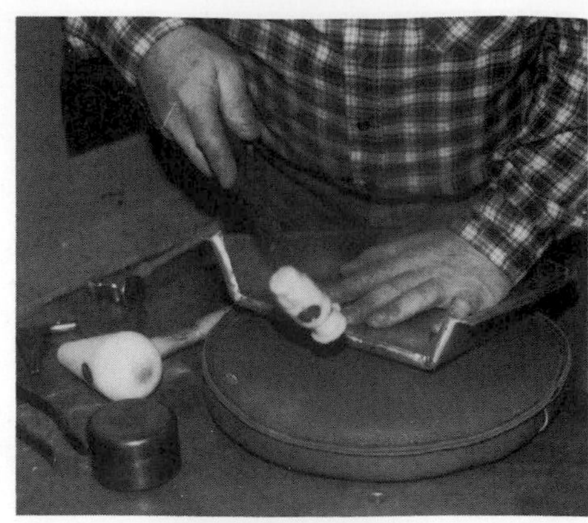

The scoop is resting on a leather shot bag so I won't scratch it. I used a teflon hammer to roll the metal the rest of the way under.

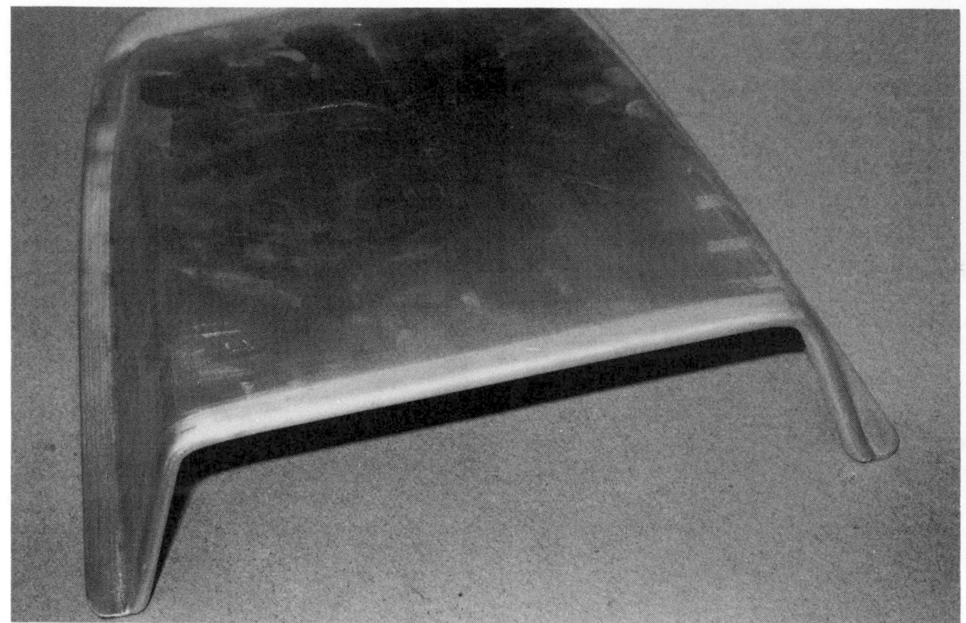

The rear of the scoop shows off the beautiful detail of the mounting flange. The graceful curved shape looks elegant and attractive. The rounded edge adds a professional appearance.

SHAPING WITH SPECIAL EQUIPMENT

There are several types of equipment designed to shape metal more easily than by hand. All of these machines perform the same stretching and shrinking functions done by hand, only they do it more quickly, with more power, more efficiently. *Eckold Kraftformers*, *English Wheels* and *Power Hammers* are all highly specialized pieces of equipment used to make compound curves. I will cover the design, function and proper usage of all of these machines. However, before you go out and use them, let me caution you: **hand-shaping metal precedes any work with this equipment**. Hand-shaping is an essential first step, and you must learn this technique reliably, before you can begin to learn to shape metal with machinery.

STATION BUCKS

The large shaping equipment I'm about to describe can easily make large parts, such as whole fenders and other body panels. These large parts will require a large form to check the shape as the metal is shaped. *Station bucks* are full-sized, three-dimensional forms of the *inside* dimensions of the part you intend to make. The station buck works like a dressmaker's dummy. It is a stationary representation of the desired shape, used to check the accuracy and fit of the part.

A station buck is not used as a base over which to form metal. It is a form to *check* the shape of the metal as it is being formed. The station buck is essential for making large, complex shapes.

A station buck is comprised of a series of sections, not a full surface representation of the finished part. Instead, each section represents a place where the dimensions of the part will change. Three types of sections are used: one vertical section, one horizontal section and several 90-degree sections, or *stations*. Stations are placed along the center section from one end to the other.

The station buck must include all the important shapes of the desired part. The metal is checked as it is formed to see if it matches the shape of the station buck.

Sometimes you can also use a station buck as a *tacking jig*—a holding device to allow accurate alignment of metal parts for tack welding. *Never use*

Station bucks are constructed in as many sizes as metal parts. This is a big one, for an air deflector on top of a truck cab.

The finished air deflector fits perfectly over the station buck. The fit guarantees the part is right.

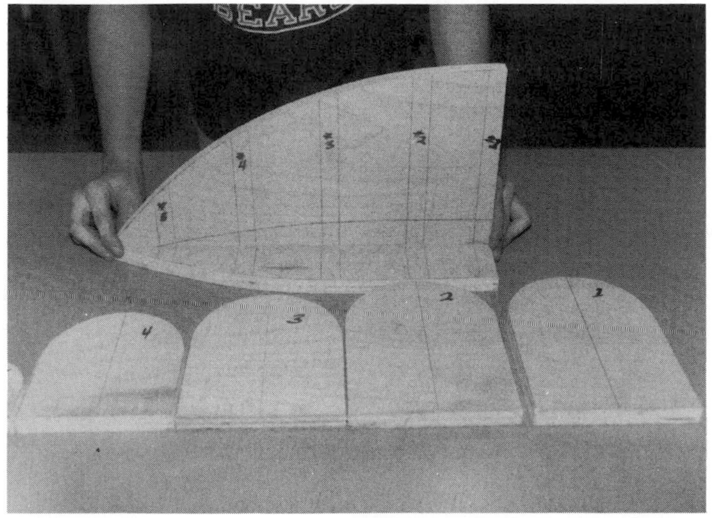

I made full size chipboard patterns for each station of the station buck. This particular buck needed seven stations.

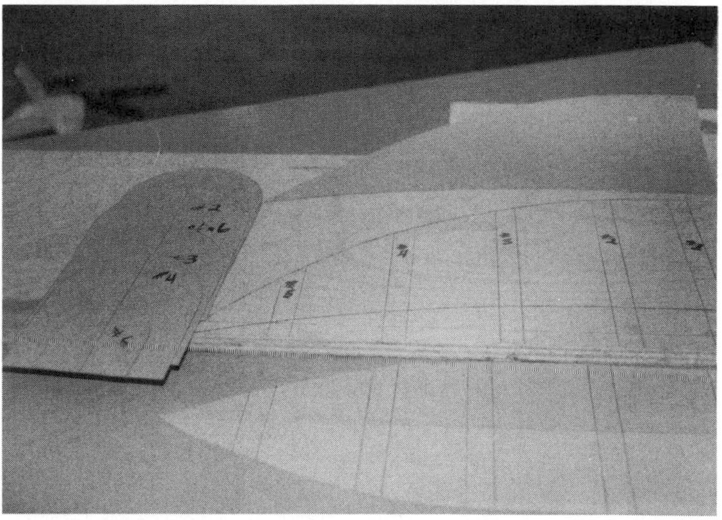

After cutting them out in a band saw, each station must be numbered or lettered to make sure it will be assembled correctly.

a station buck to hold metal for final welding. It could burn up, especially if it's made of wood.

Making A Station Buck—Begin with full-size drawings of the part you want to make. Draw both a top, or *plan* view and a side view. Make station drawings by drawing 90-degrees to the horizontal and vertical sections every 5-inches. You may need station drawings spaced closer than 5-inches if there is a major change in the surface shape—where the surface drops, rises or ends. Draw enough stations to clearly define each section of the final part. Each station represents a cross-section of the finished part.

Transfer each station drawing to 1/2- or 3/4-inch plywood. Cut out each station carefully. Number the stations, or use a letter code, to keep them in order. Provide a *horizontal base station*—a bottom surface that can be clamped to a work bench. Drill 2-inch diameter holes about 1/2-inch in from the outside edge of each station, so C-clamps can be used to hold each piece of metal to the buck for fitting.

Use wood screws and wood glue to assemble the station buck. It has to be strong and stable. Take time to make it accurate. The stations should blend together smoothly.

You can also make station bucks of steel, welding the stations together. This is a good idea when you intend to use the buck over and over again, rather than just once or twice.

It is very important for the buck to be *absolutely accurate*. If it represents the final shape you're working to get, you better be sure it is the shape you want and need. The metal part will only be as accurate as the station buck.

ECKOLD KRAFTFORMERS

Eckold is a Swiss machine tool company, marketing power equipment worldwide. Included in Eckold's fine line of metal-working equipment are three electric-powered *Kraftformers*. They code the Kraftformers by the throat depth each machine can handle, measured in millimeters.

KF 665 Kraftformer—is the largest and most sophisticated of the three, and has many interchangeable tools that give it more versatility than the other two models. With a throat depth up to 26.5-inches (665mm), the 665 stretches and shrinks aluminum and steel easily, and without marring the surface as long as LFA or LFR die sets are used. This means little or no metal finishing is required on a piece stretched or shrunken by this machine using these dies, which saves time.

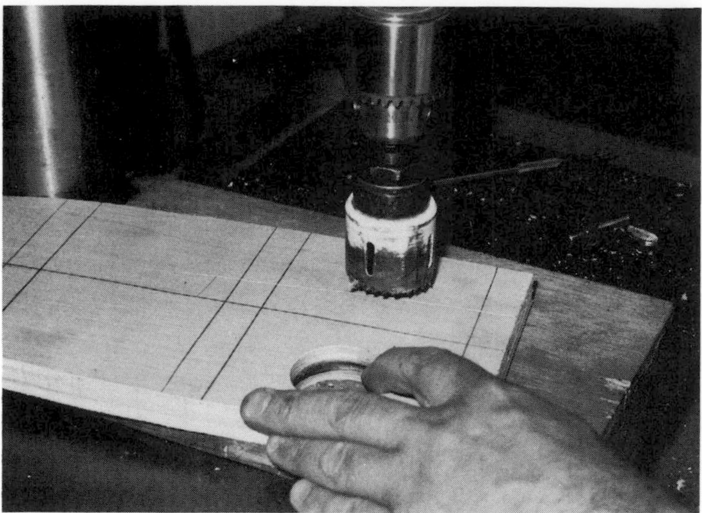

2-in. dia. holes drilled along the base of each station let you clamp metal to the buck easily. Photo by Michael Lutfy

Although the buck is now complete, I may add more holes for clamps later. As I work I'll know better where I need clamp holes to join metal sections.

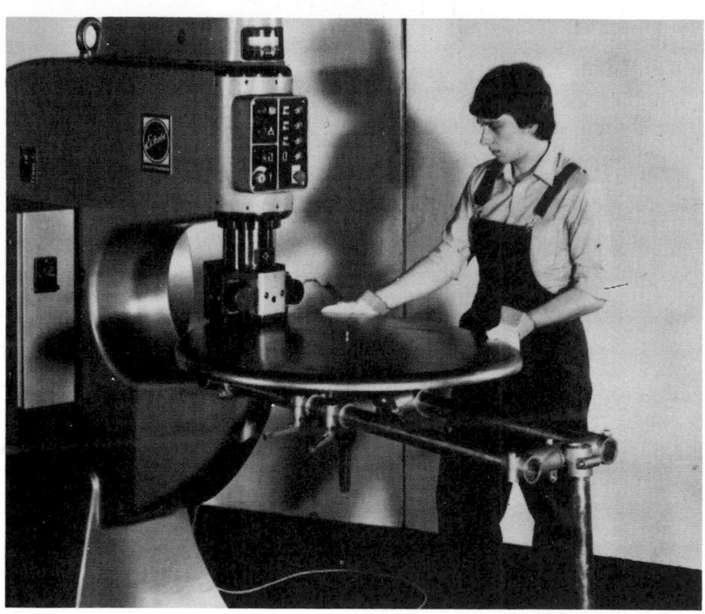

The 665 Eckold Kraftformer is the top-of-the-line model. Here it's using an edge-forming accessory. Photo courtesy W. Eckold A. G.

Beautiful, even, dome-shaped panels can be made using the PFW doming tool on the 665 Kraftformer. Photo courtesy W. Eckold A. G.

In addition, the KF 665 Kraftformer can *planish* or final-finish metal. When equipped with the PFW doming attachment, curved metal panels are possible, as well as indentations or *rises*. After using one for two years early on in my career, I was spoiled by the 665's versatility.

Eckold KF 460 Kraftformer—is smaller than the KF 665, however it shrinks and stretches metal just as well. The KF 460's throat depth is 18.1-inches (460 mm.). The KF 460 Kraftformer also has a foot control.

Eckold KF 320 Kraftformer— Recently I purchased an Eckold piccolo model KF 320. This model has a foot-control that moves the dies up and down, which allows the user to adjust the dies without having to take his hands off the work. The foot control works like a rocker. If you push down on the right, the dies move down. If you push down on the left, the dies move up. The control is very precise and easy to use.

The three die sets I use most are those for shrinking, stretching and for planishing. Most of the work I do is automotive prototype, custom or concept car work, and I've found that these three sets are all I really need. However, your needs might be different, so talk to the Eckhold rep.

You'll find Eckold Kraftformers in large automotive prototype shops like the GM tech center, and prototype facilities run by Ford and Chrysler. Two other prominent places using Kraftformers are restoration shops and custom metal shapers. A Kraftformer was hard at work when I visited Boyd Coddington's world-famous hot rod shop in Los Angeles as well.

ENGLISH WHEEL

The manufacture of auto body parts in England was once called "panel beating," a highly specialized and skilled trade. The *English wheel*, or *English raising and wheeling machine*, was developed and perfected by "the panel beaters," those who were considered the masters of this specialized trade.

Brief History—The recognition of the trade occurred circa 1900 when automotive parts were made by hammering or "beating" them out of flat sheet metal. Panel beaters would shape parts over shot or sand bags, hollowed-out wooden forms and steel stakes. This involved a great deal of hard work, because each body part had to be very smooth. As the industry developed, the trade became highly-skilled, organized and respected.

The panel beaters adopted the European craftsmen system of skilled masters training apprentices over a period of usually five years to learn the trade and become journeymen. Only after years of experience, and after proving his ability to produce excellent quality complicated panels, could a journeyman then be considered a master panel beater.

As the auto industry progressed, the panel beaters developed and perfected their methods. As cars grew larger, and competition for quality increased, so did the need to form large panels free of hammer marks and other flaws. This is the reasoning behind the invention of the English wheel.

Design & Function—The English wheel can shape steel and aluminum smoothly and easily by pushing metal back and forth using only the strength of one or two workers. No electric, pneumatic or hydraulic power is used.

The design is simple, with few moving parts. The base takes the form of a large "C", and is usually made of cast iron or fabricated steel. It is very sturdy and rigid.

The parts used to do the shaping are fastened at the open end of the frame. A flat-faced hardened steel wheel is bolted on at the top of the C. This wheel is usually 6- to 9-inches in diameter and 3-inches wide. It has an axle and two strong roller bearings for smooth rolling.

The Eckold piccolo model Kraftformer is one of my favorite tools. There is a variety of special tooling available to suit most any shaping need. Photo Courtesy W. Eckhold A.G.

Eckold makes "no mar" dies for their piccolo Kraftformer. One set lets you stretch metal without marring it, another set lets you shrink without marring. Photo courtesy W. Eckold A.G.

Mike Lewis, right, and Colin Bailey, left, are craftsmen trained in the English tradition. It takes two skilled, cooperative workers to form a large panel on an English wheel. A typical English wheel is cast iron, made in England and imported for use in metal shops in North America.

The lower wheels, called *anvils*, also have axles and roller bearings. The big difference between the anvils and the upper wheel is that the anvils are smaller in diameter than upper wheels, and they have a curved surface. The anvils are mounted in a strong steel yoke which is moved vertically by turning a large steel knob located at the bottom of the machine. This knob is attached to a large bolt. The yoke sits on top of this bolt and moves up and down with it. The yoke has a quick release mechanism to remove the metal and return to the same pressure setting easily.

USING THE ENGLISH WHEEL

An English wheel works best when you follow certain rules of operation. If you work by the rules, you'll get good results. If not, you will have problems. Remember, apprentices spent *at least* five years learning this trade. During much of those years they were learning how to use the English wheel.

First, you should remember that it takes *several different wheelings* to achieve a high-crowned curve in a metal panel. Start with a low-crowned anvil, and work up to more distinctly curved anvils as the shape progresses. If you start with an anvil with too much crown, you will mar the metal.

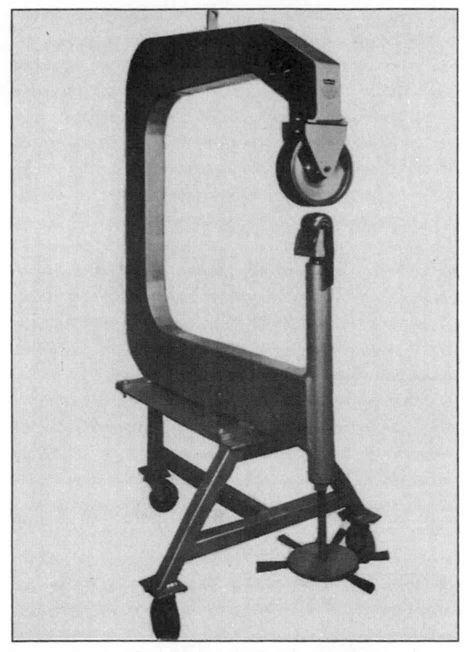

Recently U. S. Industrial Tool and Supply Co. began producing a fabricated steel English wheel in the United States. It's available with a variety of anvils. It's very strong and reasonably priced.

It is possible to make your own English wheel, such as this fine homemade example shown here.

The difference between the upper wheels and lower anvil is that the anvils are smaller in diameter and have a curved surface. Photo by Michael Lutfy

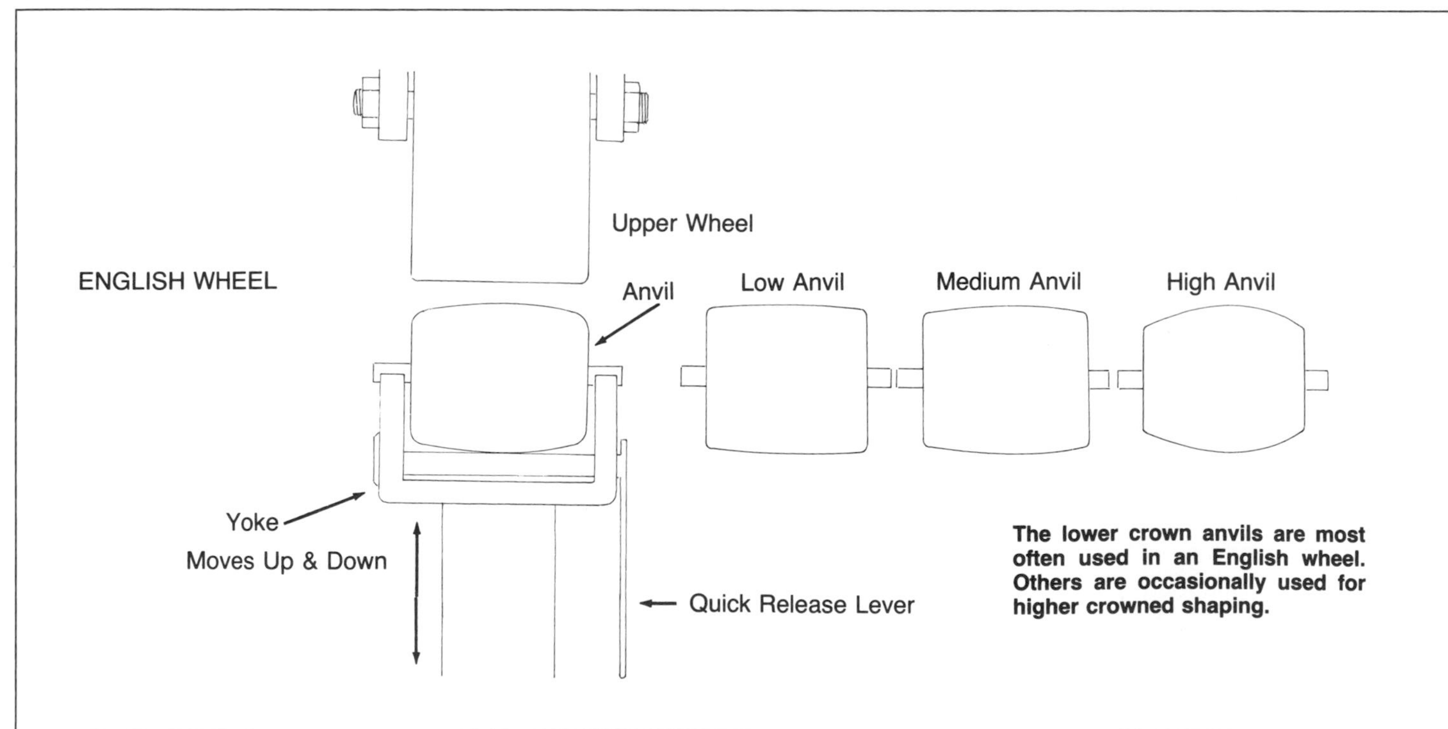

The lower crown anvils are most often used in an English wheel. Others are occasionally used for higher crowned shaping.

Always start out with a slight amount of pressure against the metal, just enough so the metal won't skip or slip through the wheels. Too much pressure will produce roller marks and mar the metal. The first few passes of the metal panel through the wheel will show if the pressure setting is correct. There should be some shaping, but definitely no marring.

To begin, insert the metal panel into the wheel and adjust the yoke. Then, move the metal panel through the wheel in a gentle, smooth, back-and-forth motion, as you guide the panel laterally through the wheel to form the curve. *How the metal is moved through the wheel is very important*! Each pass of the metal through the wheel must be close enough to the previous one so they overlap slightly. Otherwise, the metal will stretch unevenly, and the surface will be irregular. That means extra work.

The actual motion of moving the metal panel back-and-forth through the English wheel is called *tracking*. However, the panel must also be moved from right-to-left, or from left-to-right as well, without missing any

It takes several different wheelings to get a high-crowned curve. Start with light pressure and increase it as you move the metal evenly through the wheel. Photo by Michael Lutfy

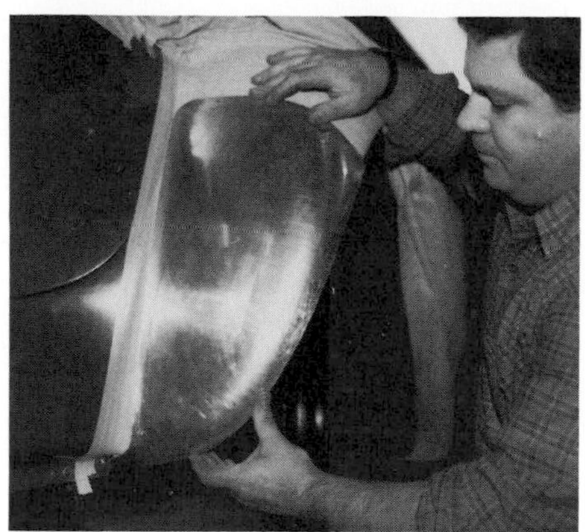

This Cobra fender guard shows what is possible once you master the English wheel.

Chicago Pneumatic's 36-in. planishing hammer is a very popular size. Unfortunately, it's also out of production. You'll have to shop for a used one.

section of metal. Each track must be parallel to the previous one. The wheels press the metal in a long line, and the idea is to move evenly so all of the metal is stretched evenly. *Don't zigzag across the panel.* This will cause waves instead of a smooth continuous curve. Tracking correctly takes a great deal of time and patience, so don't get discouraged if you fail the first time out.

If you want more shape on one side than the other, do more than one wheeling. Start out by wheeling short passes on the area you want to be curved more. Then go over the whole part and overlap the first wheeling.

AIR PLANISHING HAMMERS

Air planishing hammers smooth, or *planish* metal that has been shaped by hand. Although they are powerful enough to do some shaping, they are generally restricted to planishing. Most shaping is done on other equipment.

Design & Function—The word "planish" means: *to make smooth by hammering*. After a shape has been formed over a shot bag, it will probably need some finishing to remove hammer marks. An air planishing hammer is powered by air, so it is often referred to as a *pneumatic planishing hammer*. The top hammer head is flat, and moves up and down to do the hitting. The bottom hammer is stationary, and curved. There are several different bottom heads available with different degrees of curve. Both the top and bottom hammers are easy to remove. The top hammers are changed to get different diameters, while the bottom hammers are changed to get the right amount of curved surface.

Air planishing hammers come in several sizes, which are determined by the length of the arms holding the hammers. This measurement is also known as the *throat depth* of the hammer. Throat depth is the deepest distance you can get into the hammer without bumping into a support column on the machine. The most common sizes are 18-, 24-, and 36-inches.

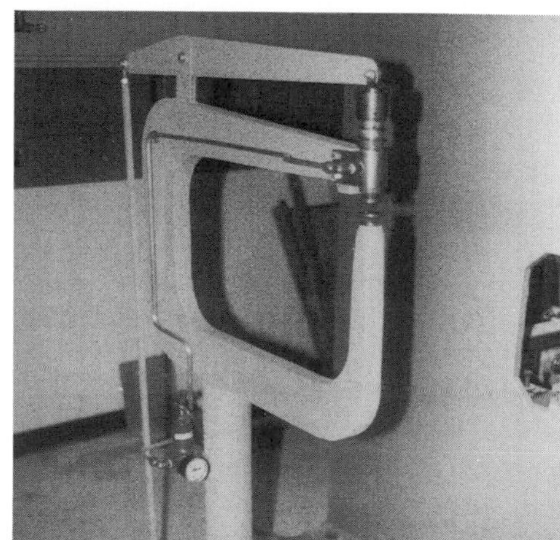

The new U. S. Industrial Tool and Supply Co. 24-in. air planishing hammer is very versatile due to its throat depth and high hammer post.

ENGLISH WHEEL TIPS

1. Practice tracking on scrap pieces of metal that are of the same alloy and thickness you intend to use most often. Start with small panels with a low crown at first, then move on to harder shapes as your skill increases.

2. It is common for beginners to put too much pressure on the panel when first starting with the wheel. This produces long bumps in the metal instead of a curve. When using the wheel remember to gradually increase the pressure settings. Keep the pressure setting the same as passes are made. Increase pressure slightly only when the curve has been completely formed at one setting, or will no longer curve at that setting.

3. Keep in mind the English wheel is primarily a stretching machine. You must be very careful not to over stretch the metal. Work slowly. Work up to the shape you want by gradual curving. Keep checking the shape with templates to see how the curve is coming along. Or use a station buck to check the curve, if you are working with one.

4. There is one last caution you must observe. Avoid running the edge of the metal through the wheel. This will stretch the edge. A stretched edge will look bad and the part won't fit correctly. A stretched edge requires the extra work of shrinking it back to shape. Shrinking isn't easy.

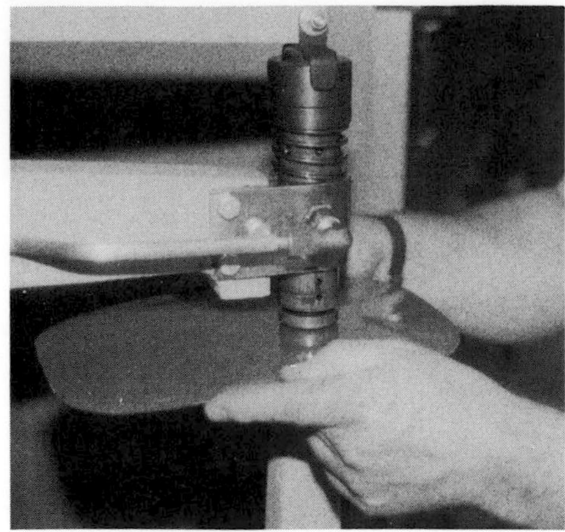

The flawless finish on this part is the result of careful use of the planishing hammer.

The lower hammer is held on a post, so metal with large curves can fit easily without striking the lower arm of the hammer. The post comes up to meet the top hammer, with a gap of approximately 1-inch between the two. The gap only exists when the machine is not in use. The large gap is there so the metal to be planished can be inserted before the machine is started. When you start the machine, the upper hammer moves down and begins to hammer the metal.

USING AIR PLANISHING HAMMERS

First, find a lower hammer head with a shape close to the shape of the metal object you want to planish. Install the lower hammer in the machine. Make sure the air line is hooked up. Now you're ready to go. Turn on the air planishing hammer depressing the foot pedal evenly.

Go slow and easy at first. Hold the metal between the hammers and press down on the pedal. It works like an accelerator on a car. Move the metal around so a large area is being hit. Don't concentrate on a small area or the metal will stretch.

The force with which the hammer strikes the metal is controlled by a small air regulator mounted directly to the hammer. Always begin planishing with a light force setting, then increase the force as the metal begins to smooth. Learn to control where and how hard the hammer hits. Planishing hammers are powerful and can quickly ruin a nice piece of metal work.

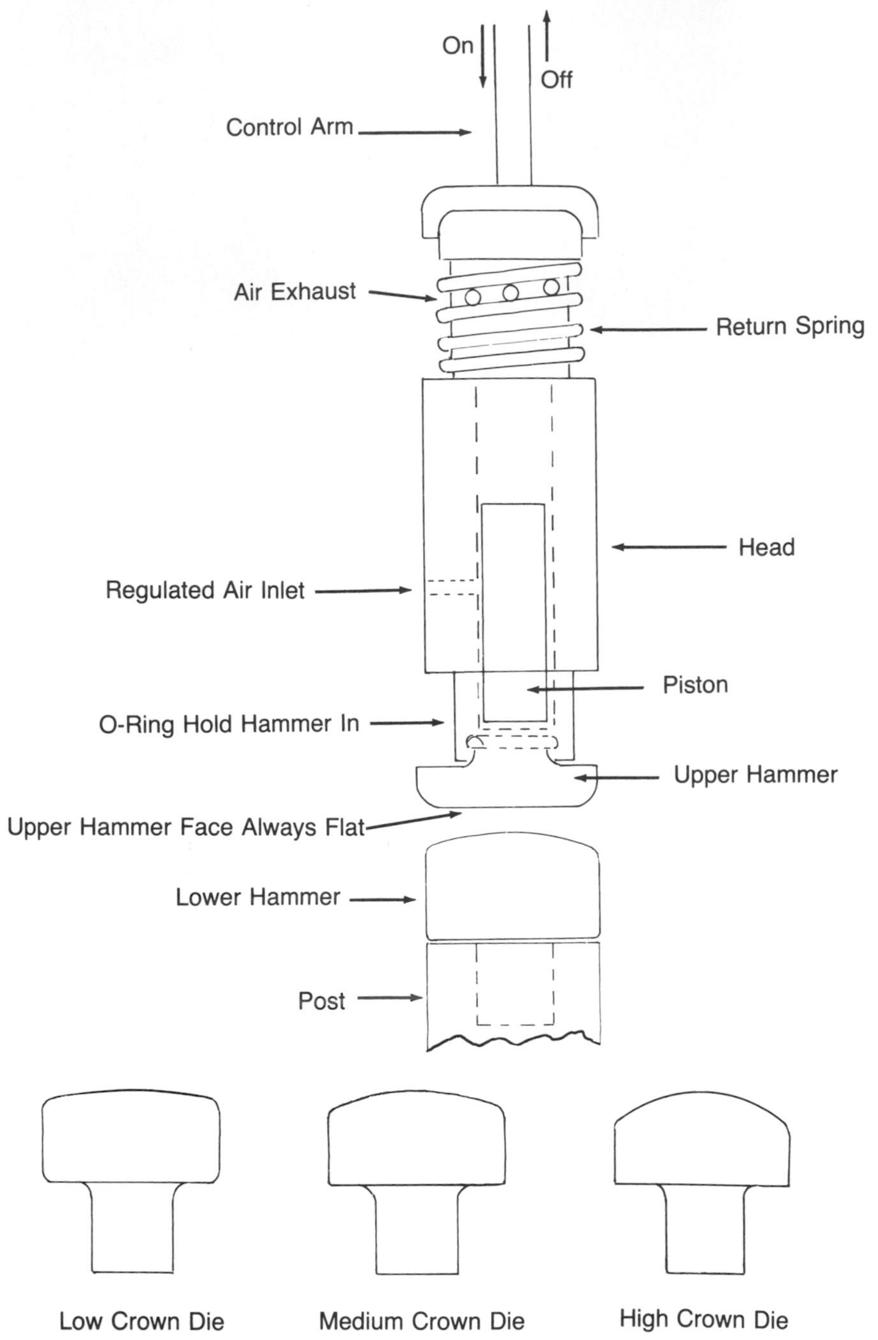

On
Off
Control Arm
Air Exhaust
Return Spring
Head
Regulated Air Inlet
Piston
O-Ring Hold Hammer In
Upper Hammer
Upper Hammer Face Always Flat
Lower Hammer
Post
Low Crown Die
Medium Crown Die
High Crown Die

An air planishing hammer is a simple mechanism. Different heads are used to planish different shapes.

As you can see by the photos, there are several kinds of planishing hammers. Most of them are sold as used equipment or surplus because they are no longer manufactured. However, a used planishing hammer is a good deal as long as it has been cared for. Check the smoothness of the upper and lower hammers and the general condition of the machine. It really helps if the ori-

USING AIR PLANISHING HAMMERS

Always be sure to keep the metal in contact with the lower hammer. In other words, hold the metal against the lower hammer at all times. If you don't, the metal will bounce between the hammers, and you'll end up with an uneven surface. The whole idea of planishing is to get a smooth, even surface.

Yoder power hammers are massive cast iron tools. They shape metal very quickly.

The low crown die set for the Yoder power hammer is in common use. This one's in use at Steve Davis' shop in Huntington Beach, California.

These are shrinking dies for the Yoder power hammer. As you feed metal in, the metal's puckered up. As you pull metal out, the raised area is shrunken by the flat surface.

ginal instruction manual is still with the machine and all the adjustments are working smoothly.

On a recent trip to California I visited several metal shops. Every metal shop I visited had one or two planishing hammers. They are beneficial to a metal shop specializing in metal shaping because they speed finish work, especially when planishing a large metal part with a deep, round form. This kind of part can be hammered out with a shot bag, then smoothed quickly with a planishing hammer.

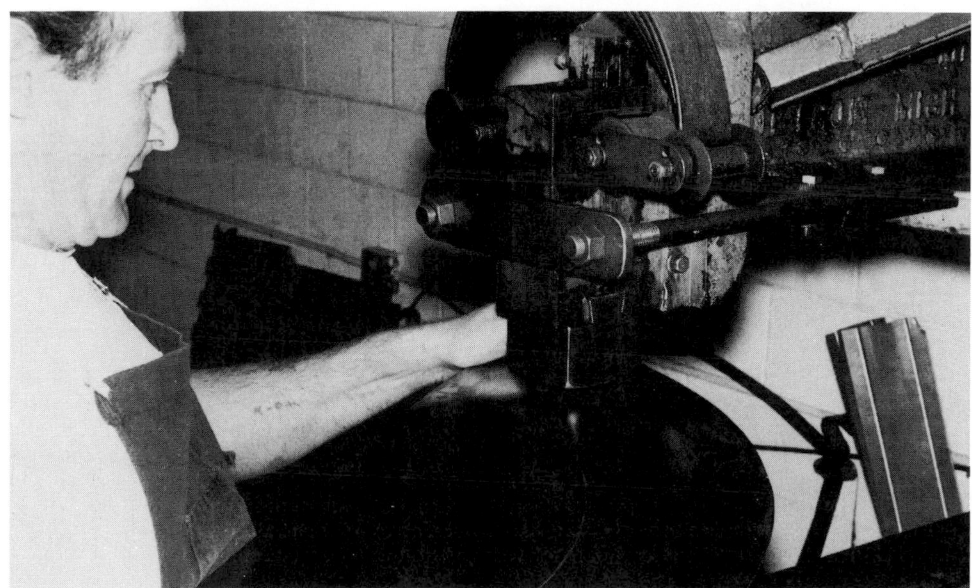

The chalk line on the panel accurately marks the depth of the first series of shrinking. Scott Knight determined the line by a paper pattern laid over a station buck.

POWER HAMMER

The *power hammer* is a very specialized piece of metal shaping equipment. Beautiful auto body parts, reproductions of antique auto body parts, race car body parts and aircraft parts can be formed quickly and precisely with a power hammer.

History—Power hammers were used as far back as the 1920s to shape automobile body parts. They were also used to form large dome shapes for bus and horse trailer tops. During the Thirties and Forties, they were used in the aircraft industry to shape aviation components, and to smooth and finish parts made with a *drop hammer*.

A person who runs a power hammer is known as a *hammerman*. Long ago, hammermen were looked upon as highly specialized craftsmen, and were separated into two general groups: large panel workers, and those who worked on smaller panels with more detail. Each type is difficult, and it was rare for any one man to specialize in both types.

The hammermen all owned and cared for their own dies. When they went to work for an employer they brought their own dies to use in the company's power hammer. The Yoder power hammer is probably the most well-known today, however it was not as popular in the earlier years. Most shops in the Twenties, Thirties and Forties had Pettingell or Trippensee power hammers. They were basically the same machine and the dies could be interchanged between them.

Design—The power hammer uses a powerful electric motor to deliver blows to the metal in rapid succession. The lower die is stationary and the upper die moves. The hammerman operates the hammer by foot control. The motor runs at a constant speed. The force of the blow is regulated by a

Scott's pulling the first shrinks. He lets the hammer touch the marked line.

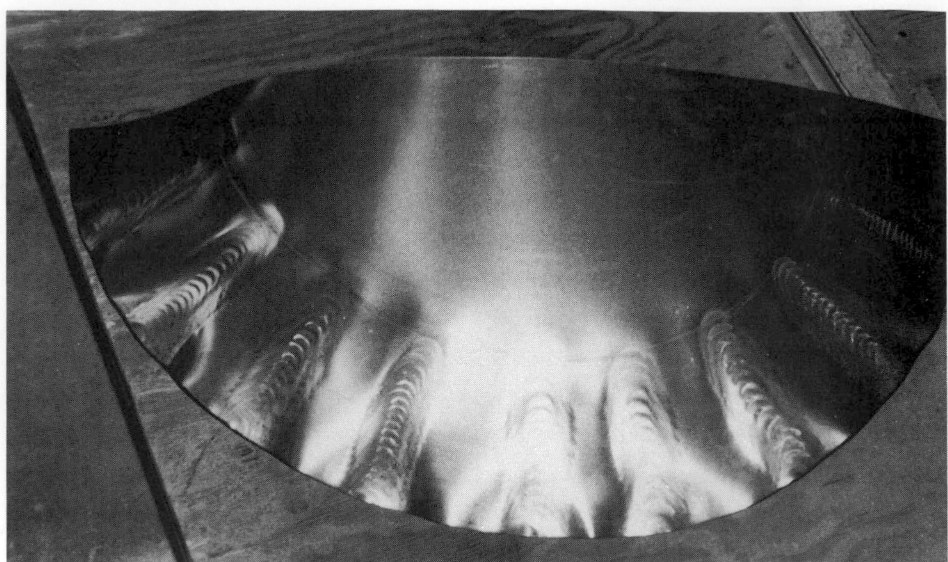

The initial number of shrinks is determined by the total amount of shrinkage required in the part. This first set of shrinks established the basic shape of the part.

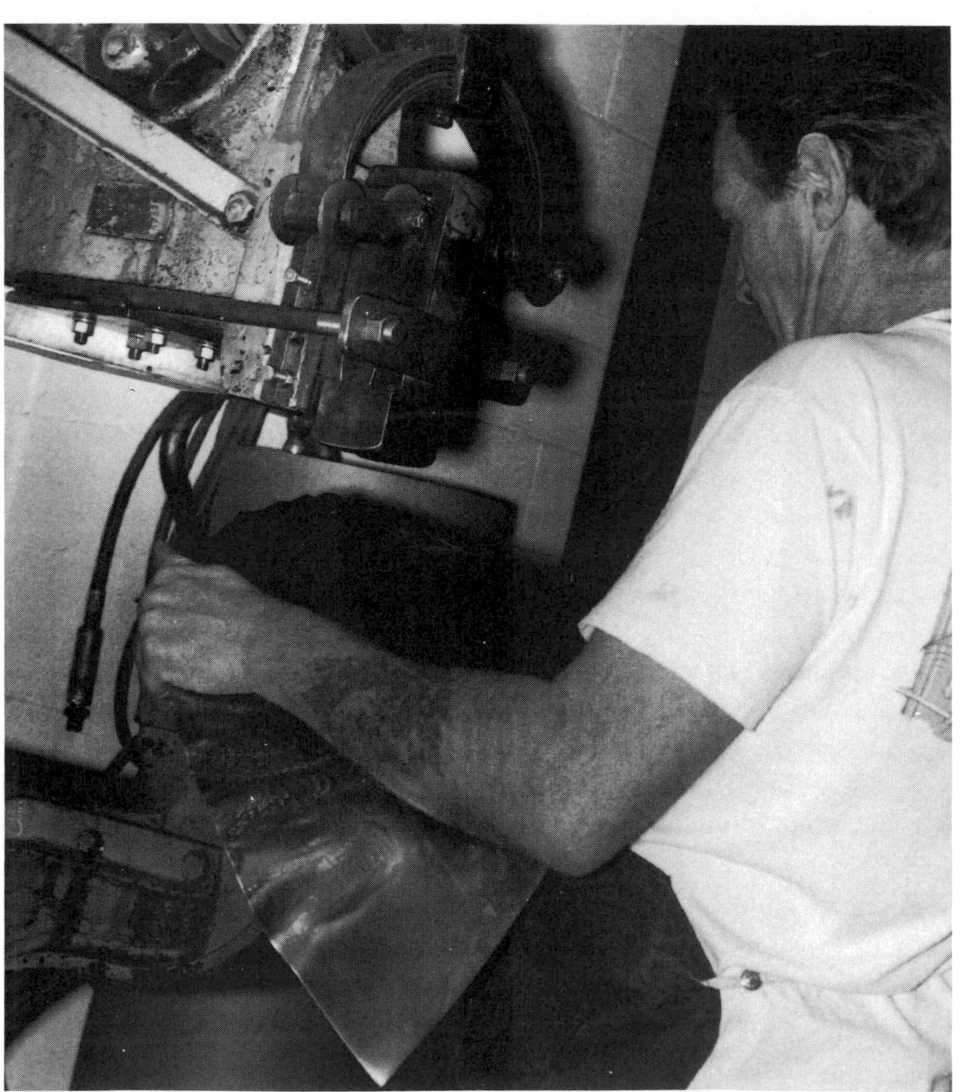

As Scott fills in the second series of shrinks he goes into the part less deeply than the first series. The *radius* of the shoulder of the part, the curve from tangent to edge, determines the depth of the second shrinks.

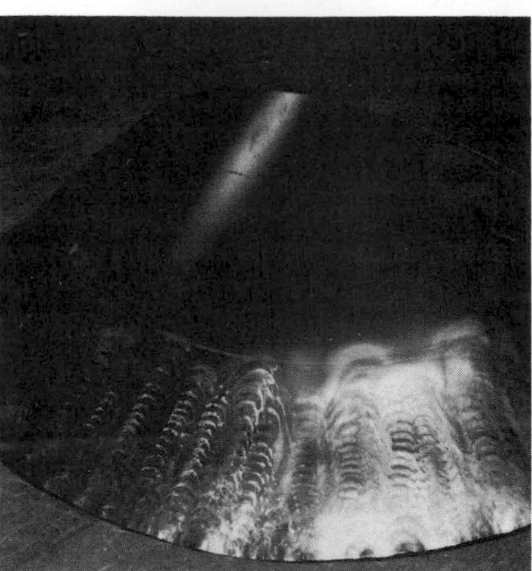

The part has been formed up. The power hammer shrinks from the inside out. This means you can fit the part against the buck as you go.

clutch connected to the foot pedal. The upper die is suspended by a leaf spring to give the blow a "slapping" action rather than a direct hit.

Dies—The dies are held by special dovetail-shaped die holders. They are locked in place with wedge-shaped keys. The key simply wedges the die into the dovetail. Changing dies is not done that often, but should they need to be, it can be done quickly.

There are five basic types of die shapes used in the power hammer.

Low Crown—The bottom die is very slightly curved. The top die is flat. The point of contact between them is nearly

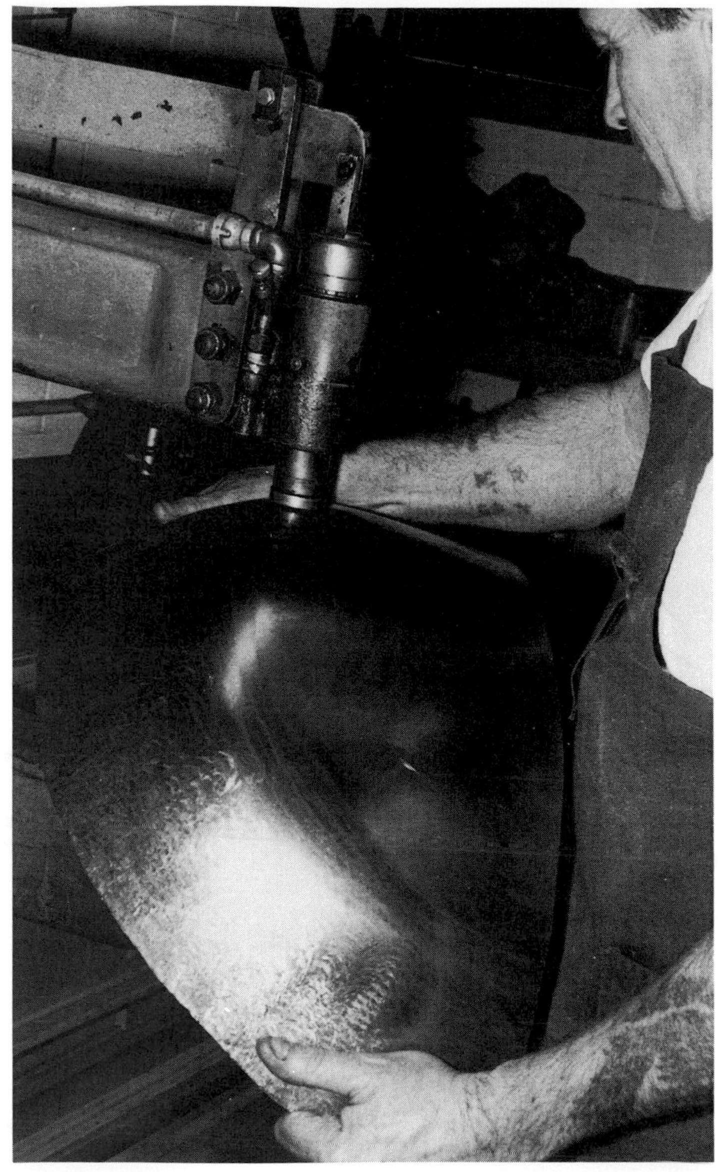

After the part fits perfectly, initial planishing can be done on the power hammer, but the finer planishing is done on the Chicago pneumatic planishing hammer. Notice how the transition from shoulder to flat is being planished.

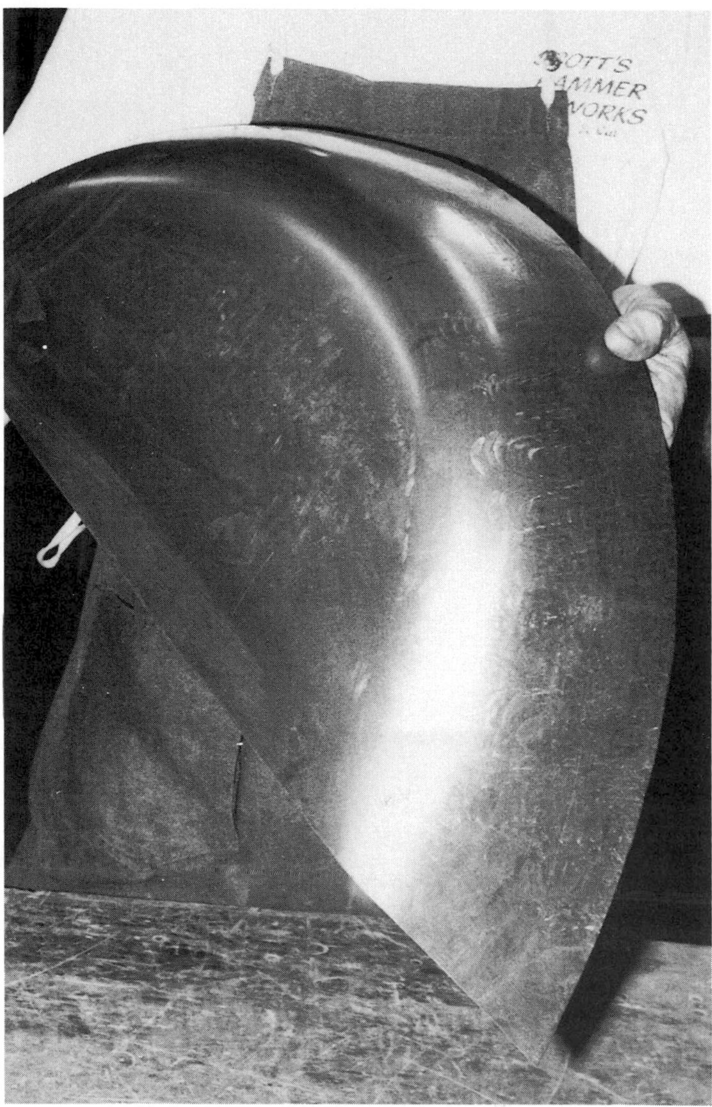

The finished part has been trimmed and can be welded to a matching part to form the whole side mount well for a Duesenberg.

the size of a quarter, approximately 3/4-inch in diameter. This die set is used for shaping low crown panels, such as for deck lids and door skins. It is also used for final planishing.

High Crown—The bottom die is highly crowned. The top die is perfectly flat. The point of contact between them is about the size of a dime. This set is used for the same kinds of shaping as the low crowned set, but it shapes faster. *It is also harder to control.* On the positive side, though, it will reach into tighter areas.

Curved Die—This is the third commonly used die set. The bottom die is curved on one axis only. The top die can be very hard rubber or steel ground to produce an oblong point of impact. It is used for truing shoulders and smoothing highlight lines.

Cross Die—The bottom die is ground to a peak along one axis. The peak runs from front-to-rear. The top die is perfectly flat. This die set is used for *stretching in one direction only*. It can also be used to form a reverse crown shape.

Shrinking Dies—are also used on a power hammer. These dies are specially-ground to gather the metal as it is moved into the hammer. As the metal is moved between the dies, it is forced upward to form a hump about 1/2-inch high and 3/4-inch wide. The raised hump is then flattened, or gathered, by the flat section of the die. This flattening occurs as the metal is drawn back out of the machine. This die set creates a situation in which it is easier for the metal to shrink when it is hit than do anything else.

In the hands of a good operator, a power hammer will out-perform any other means of metal shaping. It is the hammerman's expertise and experience that determine the quality of the part, and the time it takes to make it.

Scott is shaping the rear trunk lid for the '58 Ferrari, a panel that requires a sweeping, low-crown shape. The power hammer is both powerful and controllable enough for this task.

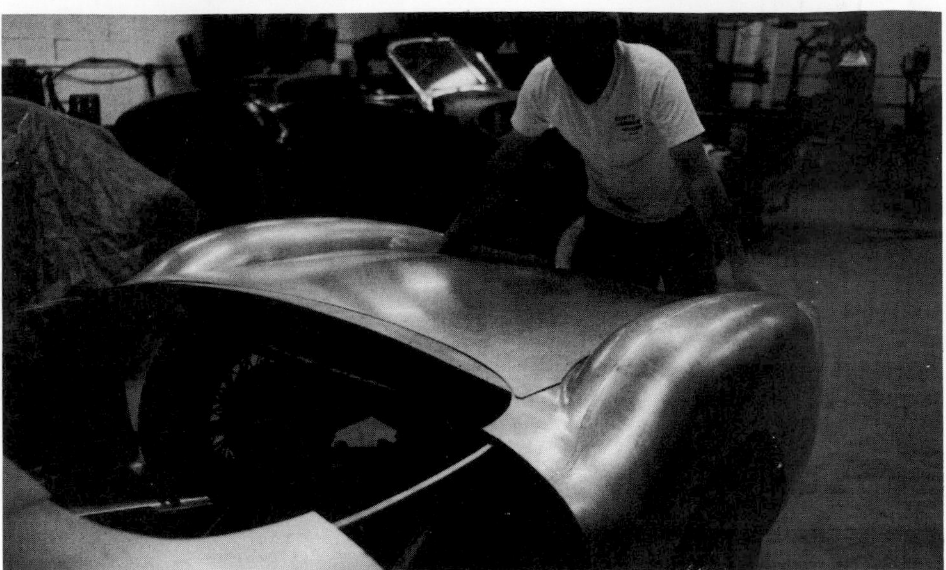

It's best to have the project nearby, because shaping with the power hammer is a gradual process that requires constant checking for a perfect fit.

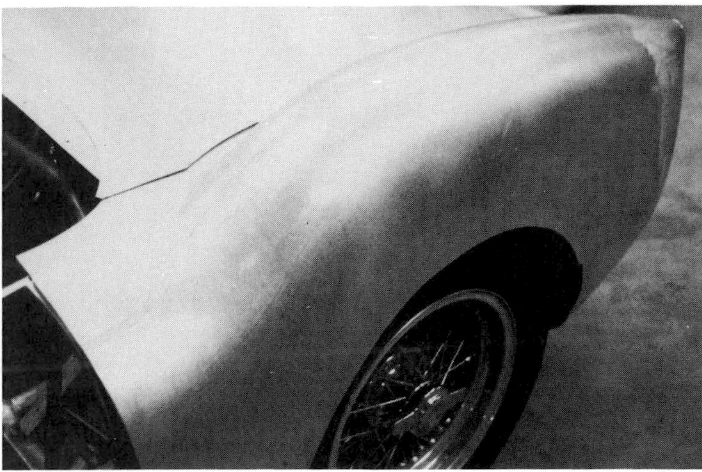

The entire tail of the Ferrari was made of several high-crown shapes formed on the power hammer.

Care, craftmanship and mastery of power hammer techniques are responsible for these fenders. They are made from individual pieces which have been accurately assembled.

POWER HAMMER POTENTIAL

These photos fully illustrate the capabilities of a power hammer when the techniques are fully mastered. Scott Knight is forming the panels for a '58 Ferrari and the fender for a '32 Ford. With a lot of practice, patience and perseverance, you could be forming your own panels like this some day!

The Yoder power hammer can shape steel easily. A high crown die set was used to form this '32 Ford fender.

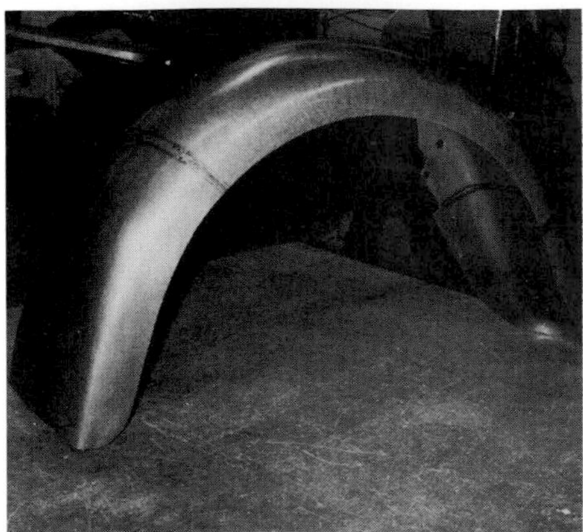

The entire fender was actually shaped from four individual pieces. Once all of the metal shaping is done, a wire is rolled into the fender edge for a finishing touch.

CHAPTER

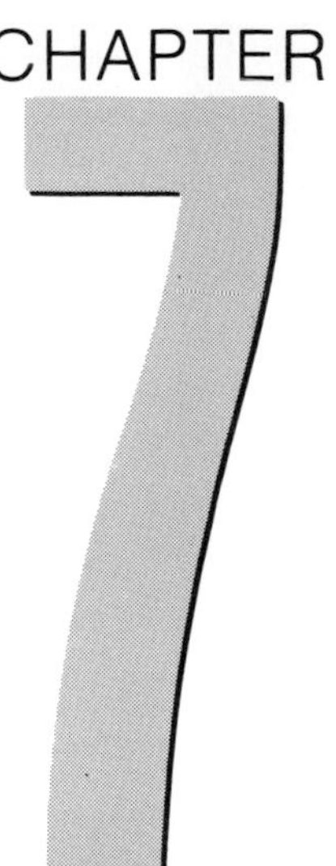

HAMMERFORMING

Hammerforming is shaping metal by hitting it with hammers, mallets or corking tools over, onto or into a base form, which is known as the *hammerform*. It may involve the use of all the elements I just described, or it may involve only some of those elements. It is an inexpensive method to make parts, whether it's a one-of-a-kind special or many of the same kind. Hammerforming is a technique available to most people because it relies more on skill than on expensive equipment. It can also produce beautiful results. Inevitably, when I show someone how to hammerform metal I get a reaction of surprise. No one seems to know how simple and effective it is.

Recently I had to make a tank end for an intercooler. The intercooler was for a one-of-a-kind car. The shape of the part was complex, and I knew it was worth making the hammerform even though I was making only one part. Hammerforms are generally more often used when I want to make several of the same part. But when a given part must be a complicated shape and must be extremely accurate, that is reason enough for a hammerform.

On the other hand, I often make round tank ends for various kinds of tanks. I will make a round end hammerform and keep it if the shape and diameter are so common it's likely I'll be making another one sooner or later. This stock of general use hammerforms can "save the day" when the need arises. If you make a hammerform of a general size and shape for one project, keep it. You may be using it again sooner than you think.

Another advantage to hammerforming is that it is possible to make parts with *symmetrically opposite halves*—or mirror images of a part that can be welded together. The result is a part that is just as strong, yet far lighter than a similar piece fabricated out of a solid block of metal. Hammerformed parts save weight without losing strength.

Hammerforms are often overlooked as a technique in sheet metal fabrication. I think this is because people think it is complicated to make a hammerform, or to use one. Sometimes they think it sounds like too much work to bother making a hammerform. Nothing could be further from the truth. I can't stress enough how simple hammerforms are to construct and that they are well worth using.

TYPES OF HAMMERFORMS

The most common hammerform is used under the metal and the metal is formed over it. In this case, the metal is being *shrunk* to produce the final part. Using the tank end as an example, the metal is placed over the hammerform, and as it is hammered upon, it shrinks down and over the form to create the tank end. This is an example of a simple hammerform.

Some hammerforms are more complicated, but not a great deal harder to use. They require metal to be both shrunk and stretched. The best example I can offer is the hammerform for a wing rib. A wing rib has both a convex and a concave shape. This needs both shrinking and stretching. The *convex* shape must be achieved by

Hammerforming is shaping metal over, onto or into a base form called the hammerform. It is an inexpensive way to produce beautiful parts. It's not as hard as it looks. Photo by Michael Lutfy

The three corking tools on the left are steel. The next three are aluminum, and the far right is wood. Photo by Michael Lutfy

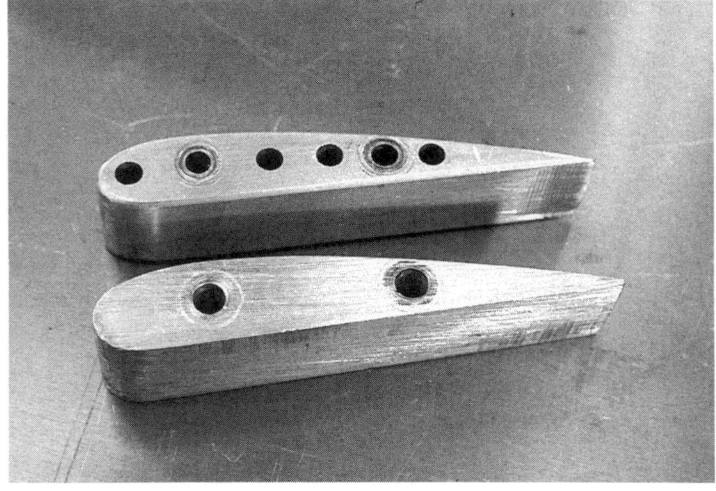

The hammerform top has six holes: two to bolt the hammerform together and four to locate lightening holes. The hammerform bottom only needs two holes for bolting. This hammerform is made of 6061 T6 aluminum.

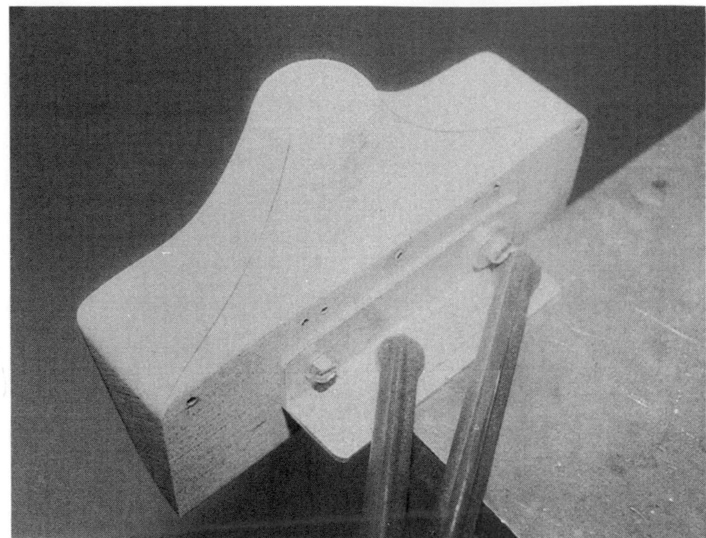

This hardwood hammerform for an intercooler tank was impossible to clamp. I had to bolt on a metal piece to clamp it securely.

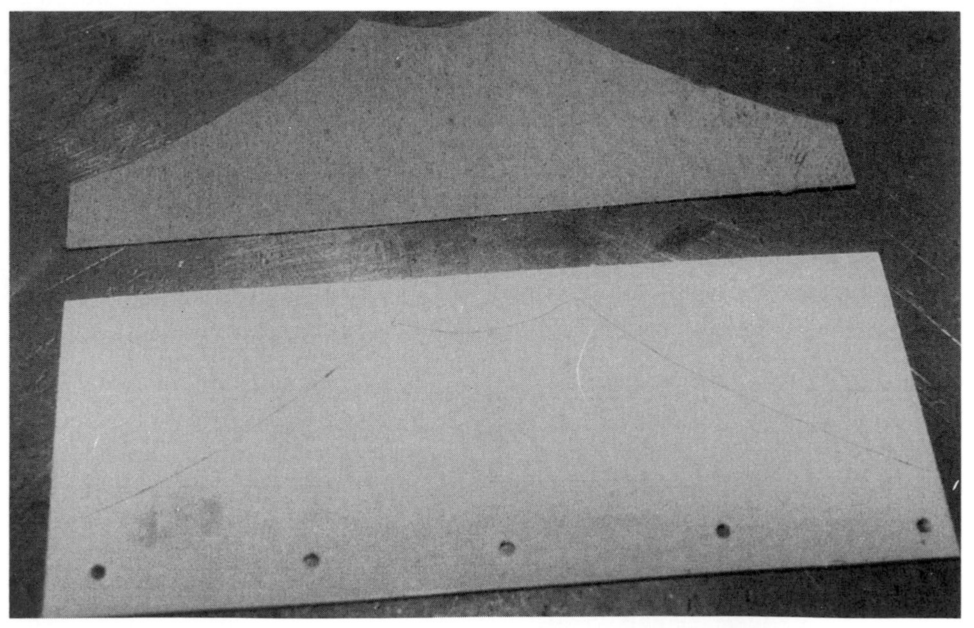

The bottom part of the intercooler tank was held down with screws because of the unusual shape. I punched screw holes in the metal blank before screwing it in place.

shrinking metal over the form. The *concave* shape must be *stretched* over the form. This combination of concave and convex shapes is also similar to techniques used for many kinds of metal sculpture.

A third kind of hammerform has a depression in the middle. You stretch the metal down into the form at the depression. This kind of hammerform requires the use of an outside locator clamp. The clamp keeps the outer shape of the metal blank from shifting. By holding the metal tight, a crisp edge can be formed on the part. The restraining locator clamp helps control the stretch of the metal into the depression in the hammerform.

A hammerform can be made from hardwood, aluminum or steel. The material used for the hammerform de-

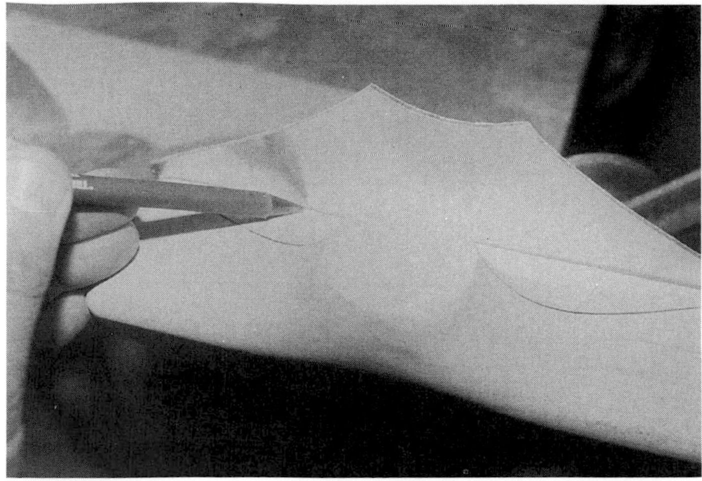

An exact bend line was marked after the intercooler tank bottom was screwed to the form.

Once the tank bottom is fastened and bent, I could begin hammerforming.

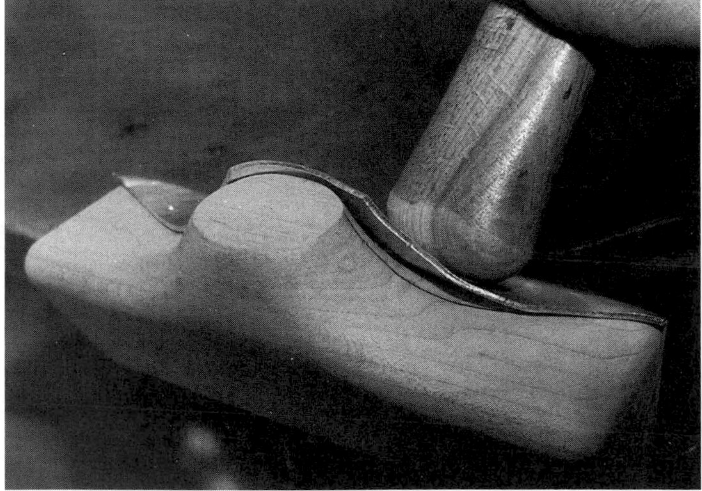

I started by forming the metal evenly toward the top.

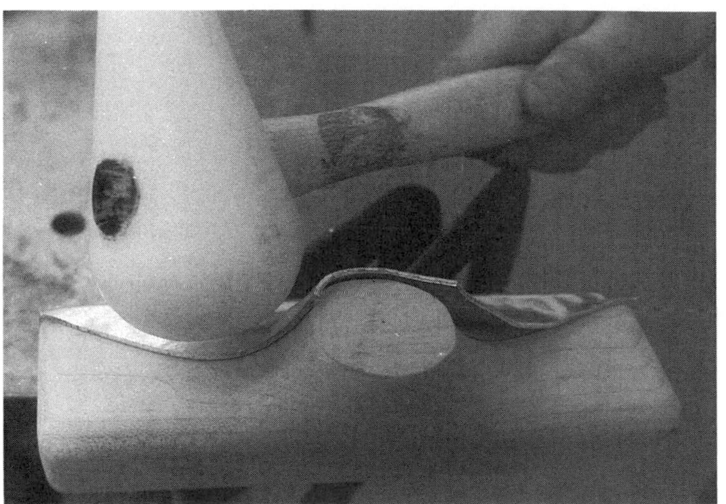

I formed the curve at the neck with a large plastic mallet.

pends on the metal to be shaped. It also depends, to a degree, on how many parts will be shaped on the hammerform. Naturally, you need a sturdier hammerform if it is to be re-used than if it is to be used only once. A project made of steel requiring many identical pieces will need a steel hammerform to stand up to repeated use.

If you intend to use a hammerform only once to make a one-of-a-kind part, make it out of hardwood. If you are using it once to make only one piece, and making the piece of aluminum, it is highly unlikely you will need anything sturdier than hardwood. Rock maple or oak are my favorites for wood hammerforms. I have sometimes used other hardwoods when they were available.

Aluminum hammerforms are made of hard aluminum; 6061 T6 or harder. Hammerforms of soft aluminum will not hold up to the pounding as metal is formed over them. The last thing a hammerform should do is distort while metal is being shaped over it.

HAMMERFORM DESIGN

Making a hammerform is a process requiring several steps. The first step is to decide the actual size and shape needed. This is where you have to be very careful. Your measurements and calculations have to be precise. The hammerform won't produce the part you need if you make mistakes at this first step, no matter how well you do all the following steps.

First, measure and determine the size and shape of the part to be made. The dimensions of the hammerform must be *reduced* by the metal thickness that will be used. If this sounds confusing, some examples may help.

Suppose you want to make a 6-inch diameter circular tank top. The metal used is 1/16-inch thick. The actual size of the hammerform would therefore be 5-7/8-inch in diameter; the part size minus the 1/8-inch on the diameter of the circle. Since the circumference edges of the part will be formed down over the hammerform, you *must* allow for the thickness of the metal itself. In this example it would be easy to make the mistake of thinking the form had to be 5-15/16-inch, forgetting the metal will be folded *all around* the diameter. Try to visualize the metal working down over the hammerform, or make a simple drawing showing how the metal will go over the hammerform, in-

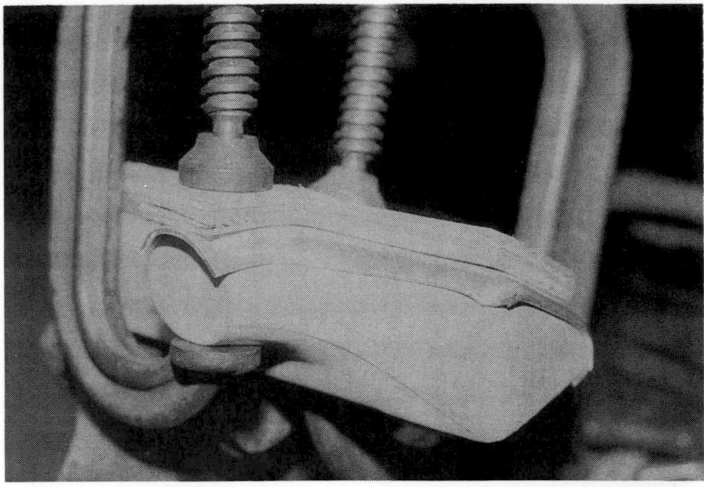

I used a 3/4-in. plywood top to hold it down, which is fairly soft and pliable, to conform to the shape of the hammerform.

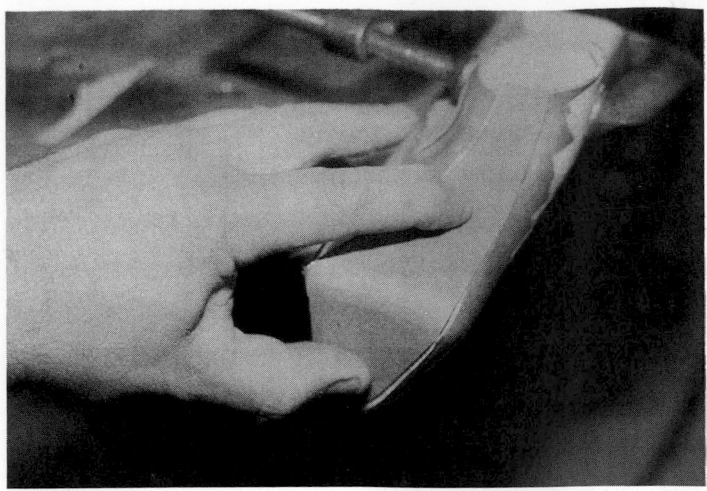

The pattern for the tank sides is fitted after the top and bottom have been hammerformed in place.

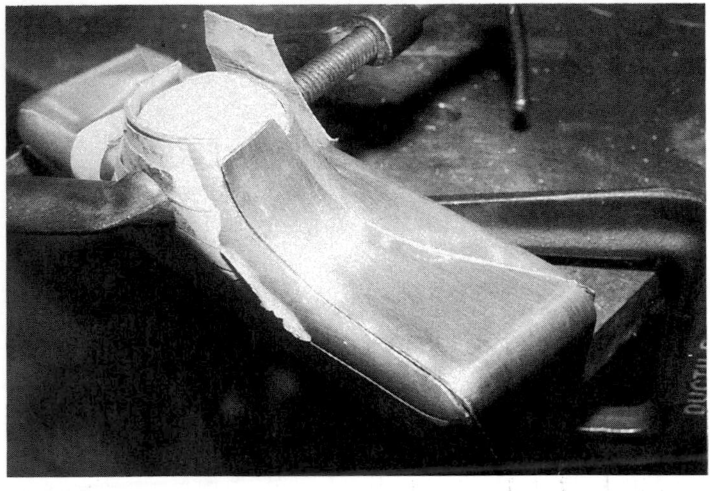

I fitted the sides very closely so I could easily make a high quality weld.

Center the aluminum blank on an aluminum hammerform. This shape is so round that a small plywood top is adequate to hold it down. This part is for the air entry on a turbo intercooler.

dicating all the metal thicknesses. It may save you from a careless error.

As another example, consider a square-shaped oil tank baffle. The baffle has to have flanges on four sides for mounting. It must fit inside a tank with exterior dimensions 8-inches wide and 12-inches high. This is a very simple hammerform. However, in order for the finished part to fit correctly, the hammerform must be *two* metal thicknesses smaller in each direction. This reduction in size allows for the accurate fit of the final part.

As a general rule, try to think of reducing the dimensions of the hammerform by twice the metal thickness to be used to make the part, or one metal thickness all around a circular part.

Before beginning, go over the drawing of the finished part. Make sure the hammerform includes all required shapes and details. Whether working from your own drawing, or no drawing at all, be sure to think through all the information about the part. Consider the complete shapes and any details that will be needed. Consider the metal type and thickness as well. If there are any doubts, write out all information and draw the shape carefully.

Once the measurement of the hammerform has been decided, and all shapes or details have been checked, it is time to make the actual hammerform, which I'll discuss in a moment. For now, I want you to be aware that a hammerform is not just a method to make parts with edges.

A hammerform can include more than a radius over which to form an outside edge. For instance, a hammerform for a wing rib may have several functions. It may have *transfer holes* drilled in to allow you to locate lightening holes in all identical parts in identical locations. The holes can also be used to locate *spars*. They can provide places to bolt the hammerforms together in order to "pinch" the work very tightly and prevent the metal from slipping or shifting while hammering.

HAMMERFORM CONSTRUCTION

Now that the measurements and special features have been decided, let's get into making the hammerform. The best assets in making the hammer-

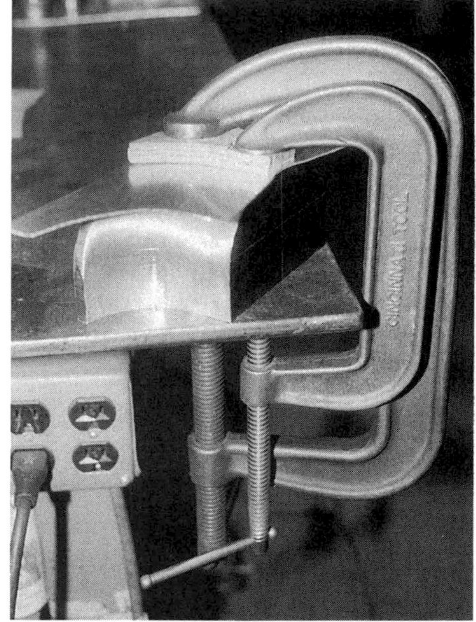

Two large clamps keep the blank from slipping. Be sure to check as you work to prevent misalliance.

I formed this part with a large rawhide mallet. The mallet conformed to the work and left no mars.

You must move the metal evenly, from the top down.

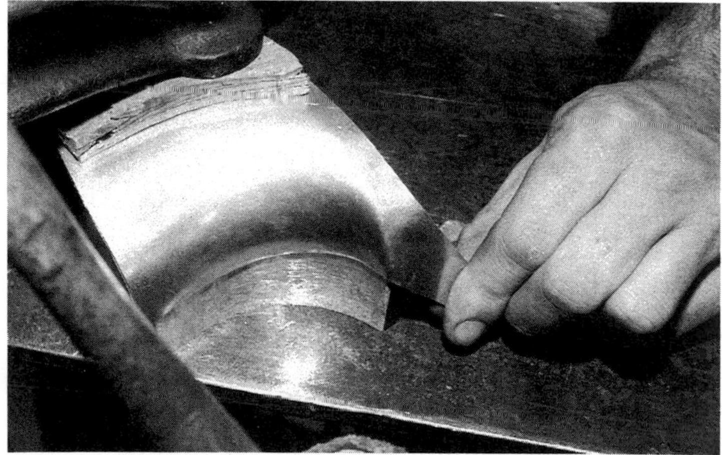

I finished forming by raising the form 1-in. from the bench with a steel riser. This let me strike low on the part without hitting the table.

form are a belt sander and band saw. If you have access to a mill, great! But I have never noticed a lack of quality when I make a hammerform with simpler equipment.

Use a coarse blade in the band saw if your hammerform is aluminum or wood. These materials can be cut faster and more efficiently with a coarse blade. A belt sander is a tremendous time saver when it comes to finishing the hammerform. *It is critical that the hammerform be very smooth, or the imperfections in the form can be transferred to the metal as it is worked.* The edges must also be radiused 1/8-inch or more. Metal formed over a sharp edge—no radius—can break or tear at the edge as it is hit.

Metal Blank—The size of the *metal blank*—a flat piece of metal which will be formed into the finished part—depends on how far or how deep the part needs to be formed. *The hammerform must be examined carefully to determine the depth of the part.* The deeper the form, the harder it will be to shape the metal blank over it. A shallow form is easier, therefore it is a good idea to learn with a piece that has shallow depth until you learn to control how the metal is formed and how it reacts. As you form along a straight edge of the hammerform you will have no trouble forming deep sections. No shrinking or stretching is necessary when the forming is straight. Almost all hammerforms will have places to shrink. Some will have places to shrink and stretch. Some hammerforms may have all three areas; straight places, shrinking places, and places that need to be stretched.

Blank Pattern—The *blank pattern* is made by laying the hammerform over a piece of thin chipboard and tracing around the outside with a pencil to get an exact outline of the shape.Cut

The finished piece is exactly 1/2 of the final part, which illustrates the "mirror image" advantage of hammerforming. I marked with a height gage where the second half will be joined. Then I trimmed off metal below the scribed line. After welding the halves together, I had the air entry for an intercooler.

Cut the hammerform, made of 6061 T6 aluminum, in the band saw. Be careful to follow your cutting line *exactly*. Any mistake will be reflected in the hammerformed part.

out the chipboard *larger* than the outline. This allows the material to fit down over the hammerform. Carefully center this first chipboard pattern over the hammerform. Hold it down so it won't slip in any direction, then push the cardboard down over the edges. Using the cardboard pattern illustrates how the metal will shape down and around the hammerform. You will be able to see how far down into the form the metal can go. You know from your drawings of the finished part just how far down into the form the metal *must* go. The trick now is to come up with a pattern producing a metal blank that will come down just far enough on the form.

At this point, the pattern can be fine-tuned by adding chipboard to alter it, or by trimming away any excess. The idea is to get the pattern to fit just the way the metal has to fit. It may take several tries, but take the time to do it right or you'll regret it later.

Lay the final chipboard pattern over the metal and mark it clearly. Cut the metal out and go over it with a file to smooth any rough edges. *This is critical if the part is going to be stretched over the hammerform.* Any rough edge, tear, scratch or imperfection may cause the metal to split as it is being stretched. The smoother the edges are before hammerforming, the greater the chances for success will be.

Lay hammerform over chipboard to develop a pattern for the metal blank. Measure and add the depth allowance for the part to the blank.

Securing The Hammerform—

The best place to do hammerforming is on a large, heavy, steel table. A light work bench or table won't work well because it won't absorb the impact of the hammering. A lightweight table might actually move around if it isn't bolted down. Work on a sturdy steel table that is secure.

The next step is to secure the metal blank to the hammerform. The blank *must* be centered perfectly over the hammerform or the depth of the part will be uneven. You might be making scrap if the blank isn't exactly in place before you put the top on.

Use another piece of wood, steel or aluminum to help "sandwich" the blank onto the hammerform. If the hammerform is made of steel, the top material should also be steel. The same holds true for aluminum or wood hammerforms. The same material used for the hammerform bottom is the right one for the top as well.

The top generally is the same shape as the hammerform base, but the sides are straight and it is smaller by 1/2-inch or more. The sides of the top piece have no radius. If the hammerform bottom has rounded sides, the top should only come to the point where the curve of the hammerform bottom begins. The top piece should be made this way so the metal can be tightly formed over the curved edge below

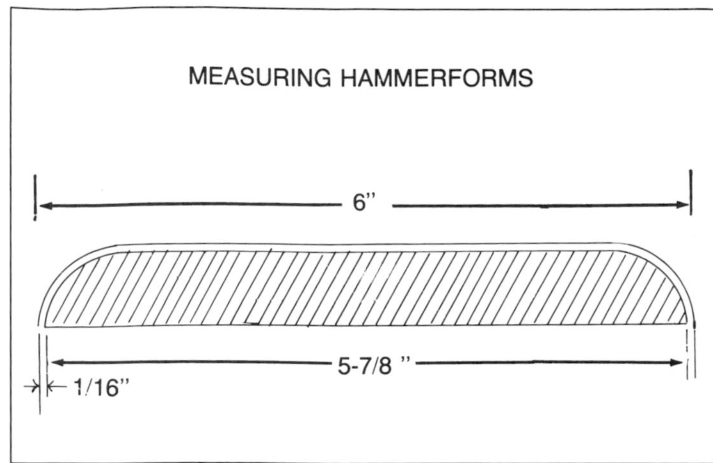

Reduce the diameter of a round hammerform by two times metal thickness when forming the metal over the hammerform.

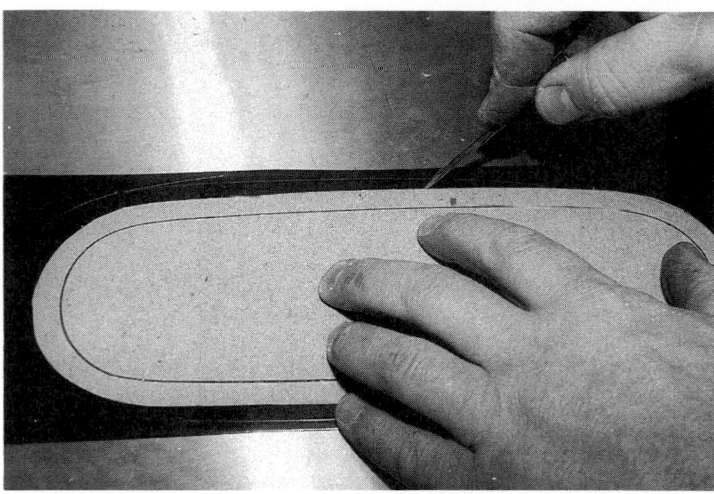

Use a scribe to clearly mark the blank all around the blank pattern.

Be *very careful* to align the stack of hammerform bottom, blank and top. Any misalignment will make the side depth uneven.

without obstructing your vision.

The top must be strong enough to hold up to the pressure of the clamps that will be used. The whole "sandwich"—hammerform, metal blank, and top—must be clamped *tightly* before forming the metal. The actual hammering during forming could make the metal blank move. *If the blank shifts slightly the part will not be accurate*. The more clamps you can use, and the tighter you can secure the assembly, the better.

Sometimes, a small hammerform can be secured in a large vise. This is a good way to hold small hammerforms to make small parts. It is very tight, secure and slipping is almost impossible. Clamps can be used with the vise, if the vise is not wide enough. Just remember to secure the whole assembly as tightly as possible.

With some projects, it is possible to *bolt* the top, blank and hammerform together. A rear wing rib for an Indy-type race car is a good example. Wing ribs usually contain holes in their design, which allows for bolts through the top, blank and hammerform. Two bolts should guarantee alignment in most cases, but sometimes, as with the case of a large wing rib, bolts, clamps and perhaps a vise will be necessary.

If you suspect the metal blank might have shifted during clamping, stop and check it. This is a critical time and shifting can easily happen. Take time to double-check before going on. Readjust, reassemble and reclamp if necessary. You'll need lots of heavy duty clamps to do the job right.

WORKING WITH A HAMMERFORM

After you have constructed the "sandwich" of hammerform bottom, metal blank and hammerform top—and securely clamped or bolted it together—you are ready to begin forming the metal blank. At this point, consider this first effort a trial run. The first time you try working on the hammerform you may discover some things you hadn't suspected before.

When the alignment was perfect, I used many clamps to secure it. I chose these three corking tools to form the part.

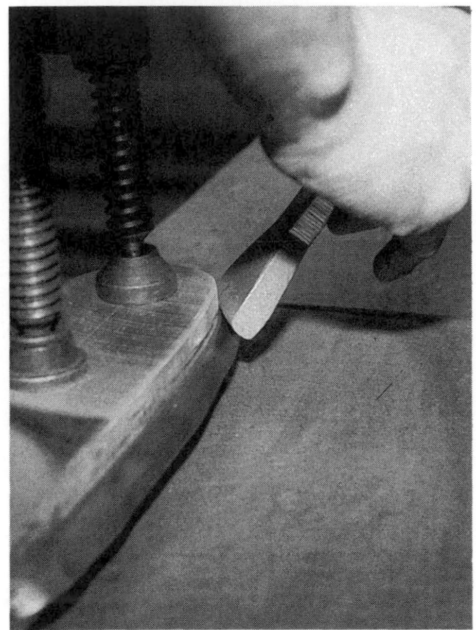

Begin forming by striking high on the part with a corking tool. Move down slowly, directing the force toward the hammerform.

The metal blank may be too long or too short. Some areas of the hammerform may be too deep to be worked, or you might find you can work more deeply than originally planned. This information can only be realized one way: try it and see what happens. I don't mind scrapping the first piece I work on a hammerform if it means I get the information I need.

It is easy to change the pattern for the metal blank if I need to alter the shape or dimensions. It can be a little more tricky to reshape the hammerform, but it's better than producing an inaccurate part. Take time for a trial run. If you are lucky or skilled enough to get perfection the first time, fine. If not, welcome to the human race.

Corking Tools—Though the metal blank can be hit directly with a hammer to form it, most of the time you will need *corking tools*. Corking tools allow you to form the metal with more control. The blow from the hammer is transferred to the metal via the corking tool. By maneuvering the tool, you control the direction of the blow.

Most of my corking tools are made of hardwood and are from 6 to 9-inches long. Hardwood is an excellent material for hammerforming aluminum. The end striking the metal can be as wide as 2-inches or smaller, depending on the required shape. The upper end should be no smaller than 1-inch square in order to hold the corking tool firmly. I have found that if a corking tool is shorter than 6-inches, it is easy to end up hammering your hand. A shorter corking tool can also obstruct your vision. Make the corking tool long enough to handle easily and to give you a clear view of your work. On the other hand, I have found corking tools longer than 9-inches extremely awkward to handle. So stay between 6- and 9-inches for corking tool length.

I keep a good stock of corking tools that can be adapted to fit a particular project. The shaping ends must be very smooth and must match the curve needed. I frequently stop working at the hammerform and "re-dress" a corking tool. I may reform the corking tool many times in the course of using it on a single job. A disadvantage with wood corking tools is that they will occasionally split. If it does split, make a new one. Don't keep hammering on a split or cracked corking tool. It may break and cause an injury, or it can mar the metal blank.

A combination belt/disc sander is ideal for redressing corking tools. The shape for the corking tool face can be taken right from the hammerform. No two hammerforms are exactly alike so you will end up with different faces on corking tools.

Aluminum corking tools work well too. They are more durable than wood and do not split, as wood sometimes does. An aluminum corking tool is best to use when hammerforming steel. They will also last longer.

Hammering—I use a medium-weight hammer, about 12-ounces, to strike a corking tool. A lighter hammer just doesn't have the force to shape the metal—especially if the blank is steel. A heavy hammer sounds like it might be better, but it will wear you out fast, even if you have muscular arms. Think of it this way: you may have to hit the corking tool hundreds of times before the part is done.

Begin hammering on the corking tool to form the metal over the hammerform. *The metal must be brought down evenly by hitting all around the blank, a little at a time.* Do not concentrate on just one area and then go on to another area. This is easy to do, because almost all hammerforms have easy spots that will tempt you to keep hammering on.

The main reason for bringing the metal down evenly all around is to avoid a potential problem. Working all around the blank evenly prevents forming a spot that can't be formed down easily. Concentrating on one spot will most likely create a bulge or *ribbon* on the metal blank. When this happens, it is very hard to work out.

If these difficult bulges or ribbons occur, they must be shrunk by hitting them with firm blows directed down and toward the hammerform. Start at the top and work down, over and over, *always concentrating on the high spots*. The low spots will eventually take care of themselves.

Throughout the hammerforming process, remember many small blows are better than a few hard ones. Work gradually and evenly all the time. Rushing and hammering hard will produce amateurish marks, and the blank will probably not fit tightly on the hammerform.

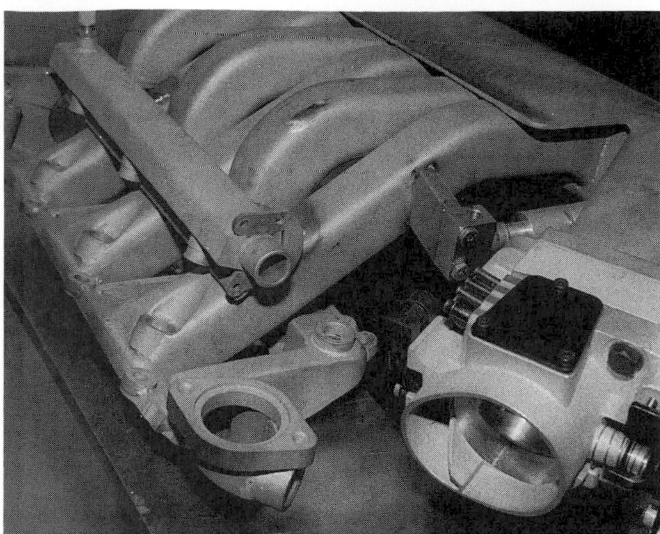

This V-6 intake manifold had two different shapes of runners. Each runner needed two hammerforms, one for the right side and one for the left.

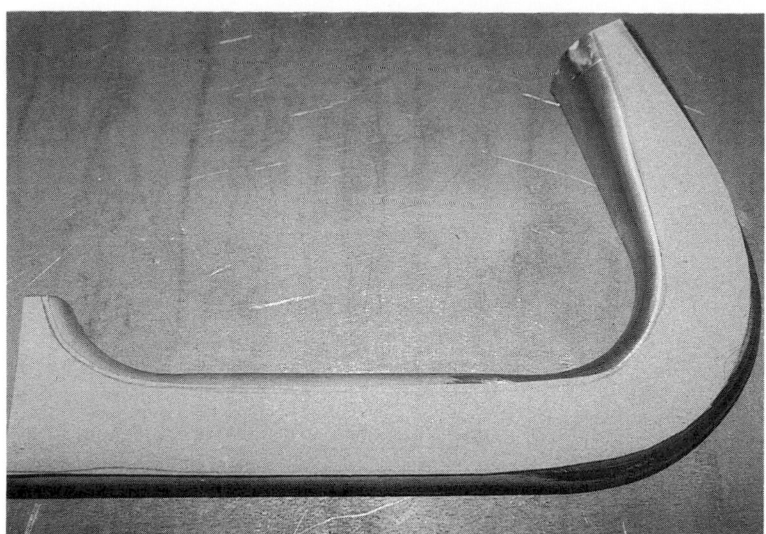

The runners were formed from 0.050-in. mild steel. This material required a very strong, solid steel hammerform.

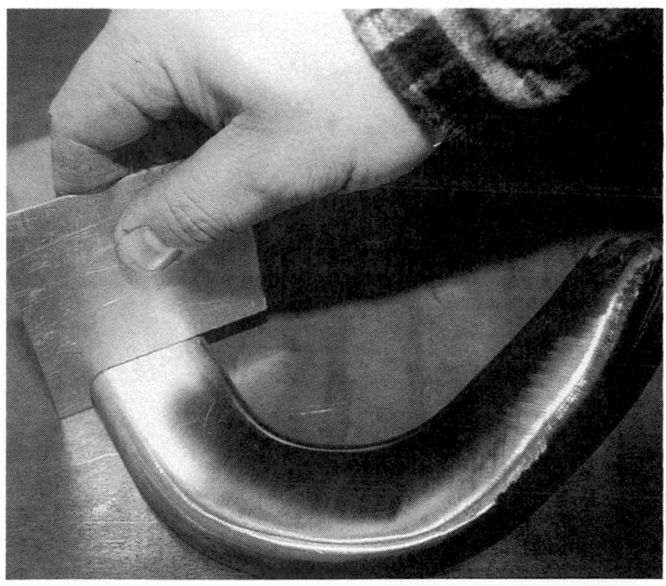

I made aluminum templates to make sure each hammerform had the right shape at the right place. All the measurements and other information came from the blueprints supplied by the factory engineers.

This pattern for the top hold down strip goes the full length of the part, but only up to the tangent. It's made this way so I can form the metal from where the bend begins.

BUILDING A PROTOTYPE MANIFOLD

Recently, I was asked to build a prototype manifold for a new V-6 engine. The engineers had designed it as an aluminum casting, but found out the cost and lead time of the casting were too great to fit their deadlines and budget. They asked if I could fabricate a steel manifold. After studying the drawings, I decided it was an ideal project for hammerforming. This sequence will illustrate just what is possible once the techniques of hammerforming have been mastered.

I realized the manifold would involve other methods of fabrication, including sheet metal layout, bending the plenum area with a combination brake and machining for final preci-

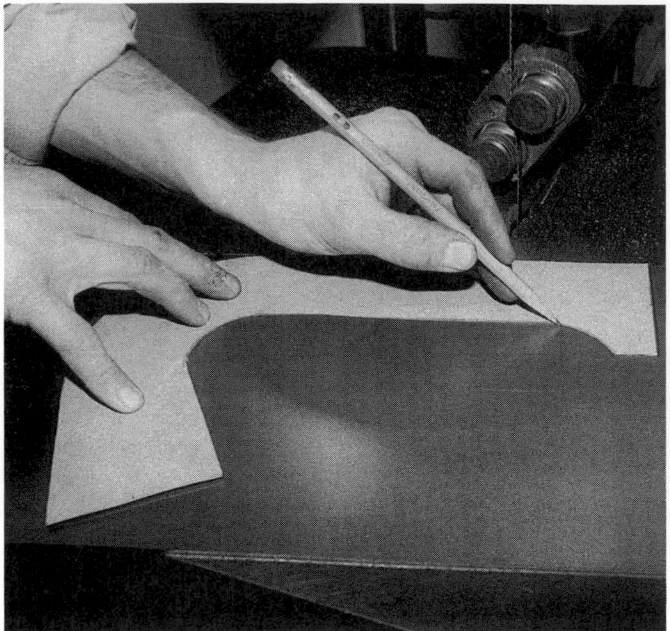

I used a chipboard pattern to make a trial part. If it formed okay I'd use it for a finished piece. If not, I'd remodel the pattern by adding to it or trimming it.

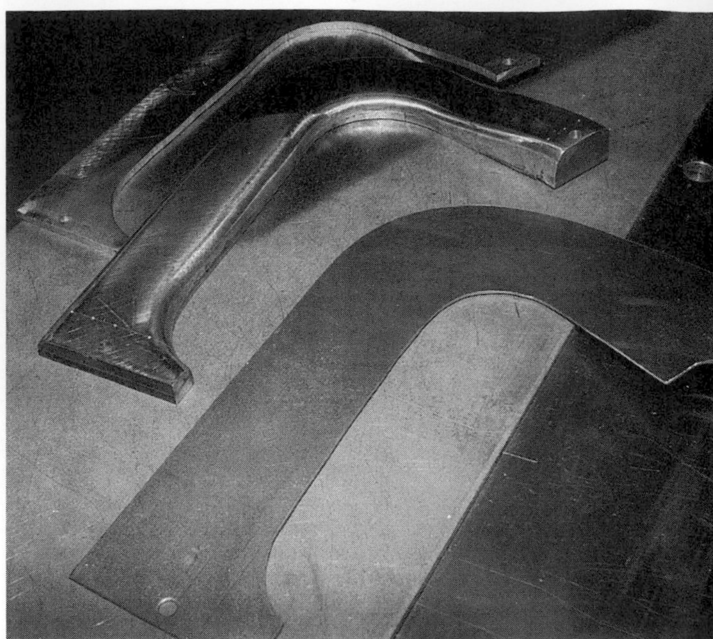

The finished blank, hammerform bottom and top are ready. The holes position locating pins, which keep the blank and top from shifting while I hammer over the hammerform.

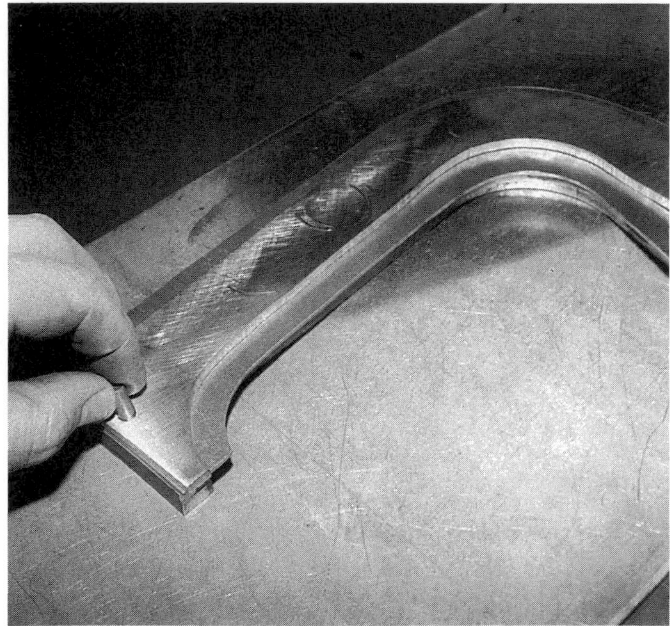

I drilled the hammerform top, blank and bottom all at once to keep perfect alignment for placing a standard 1/4-in. x 5/8-in. steel dowel locating pin.

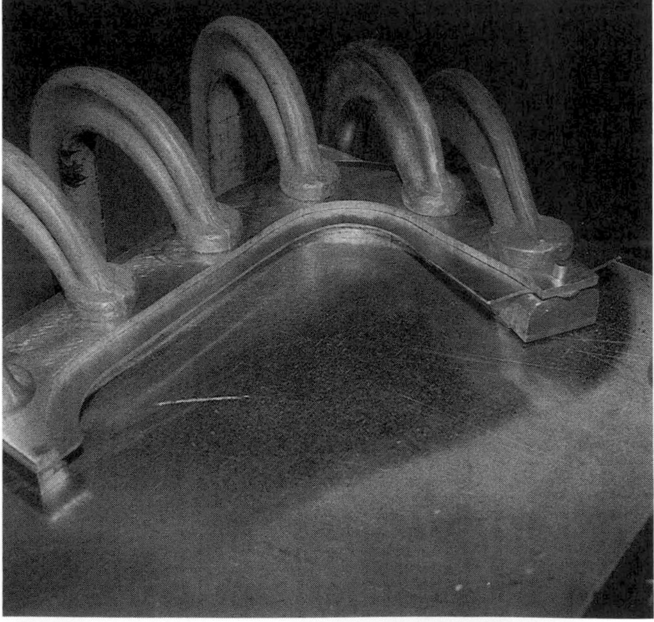

Use lots of clamps for to hold down the hammerform securely. When one side of the part is formed, unclamp it, turn the form around and work on the other side.

sion surfaces. These operations were easily within the capability of our shop.

The engineers indicated up front that if the manifold performed well on the dyno, they would want more. With this knowledge, I chose to make hammerforms from solid steel so they could be used repeatedly if necessary.

The whole manifold was composed of several different parts. I decided the plenum fuel rail and throttle body neck could be formed on the combination brake. I saved that work for last, because those parts could only be fitted

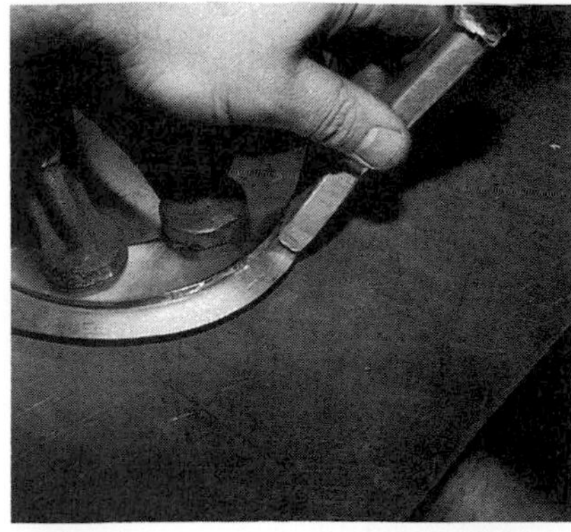

Always start work high on the hammerform to get a tight fit. Move evenly around as you form the metal blank down. Don't get one section ahead of another.

The corking tool shape must match the shape you want to form. If it doesn't, reshape the corking tools on a belt sander or grinder, depending on what material it is made of. Move across and down evenly until the metal is tight on the form.

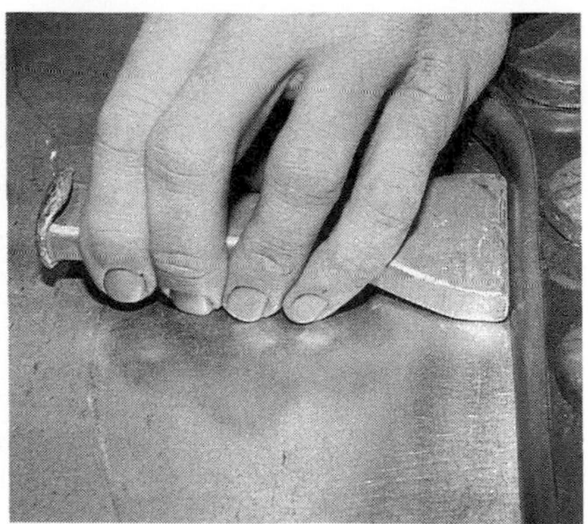

This long, straight area needs a wide, flat corking tool to keep it smooth as it is formed.

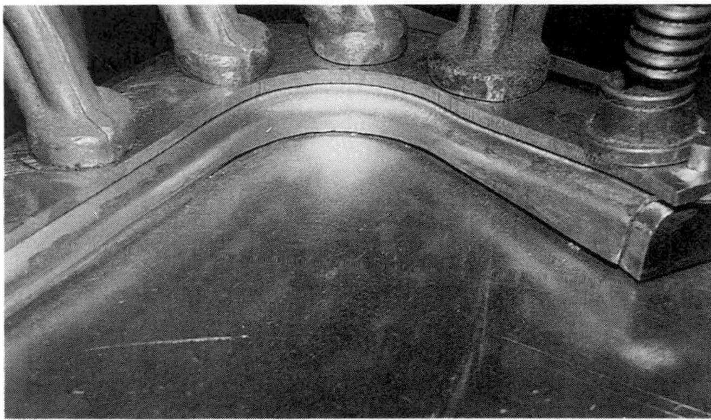

The final hammerformed part has no marks, dents or scratches. It fits the form so tightly it has to be pried off the hammerform bottom with a screwdriver.

I marked the excess metal with a surface gage, then trimmed it off. The formed half can now be joined to the other half with a strip of metal between them.

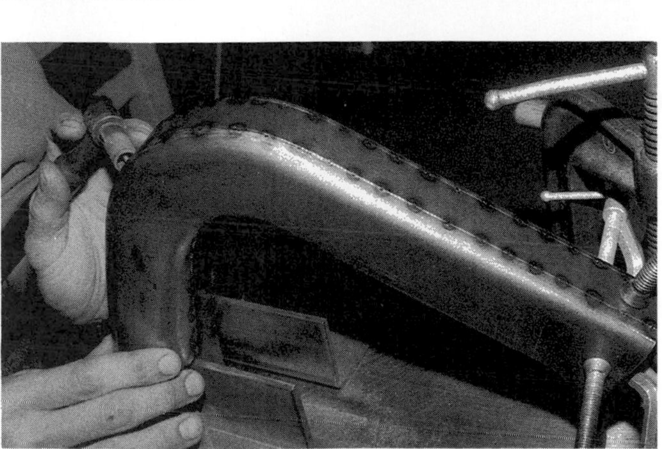

A metal strip is cut to the correct width, then bent to fit the curve of the runner. Finally, it is tack welded into position. I then filed the tack welds flush with the surface. Don't final-weld until every joint is even.

after the hammerforming and machining steps were completed.

The manifold base plates, thermostat housing flange, throttle body flange, all bosses and threaded fittings needed to be machined. I simply pulled the specifications for those parts from the blueprint and assigned their fabrication to the shop's machinist.

Once the machinist began his work, I concentrated on hand-fabricating the intake runners, the plenum, throttle-body neck and manifold water cross-over. Only the intake runners needed to be hammerformed. However, those

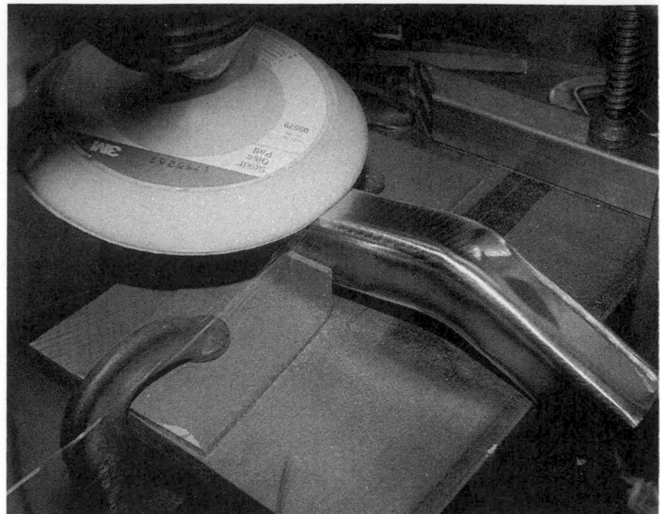

Metal-finish after final welding. Any high spots of the weld are filed off. I used a soft pad sanding disk for the final sanding. This does a nice job and it works fast.

The runner flanges are held in a V-shaped fixture. I used a side fixture to hold the runners at the right height.

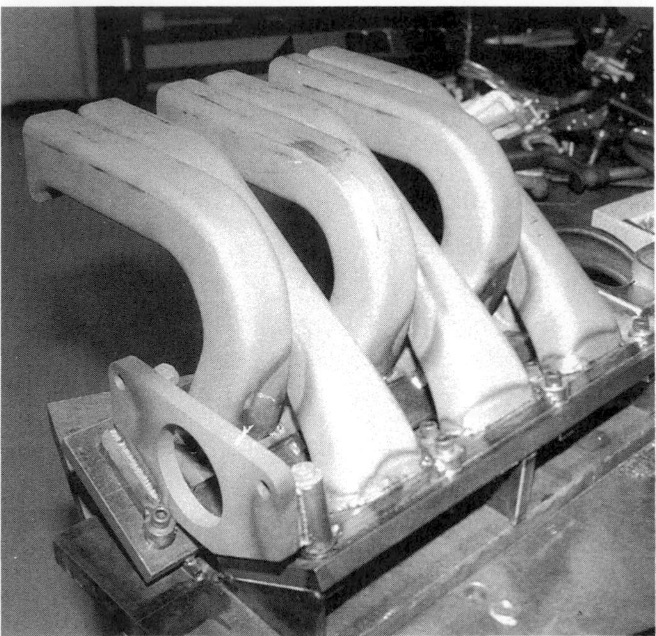

All six runners are final-welded to the manifold base.

I put the plenum on in two pieces, a top and side (arrows) so it could be welded to the inlet ends of the runners.

parts were the most important components to the entire manifold. They required a very specific shape to provide maximum airflow and proper length, according to the design.

I made each of the six runners from four pieces of metal: a right side, left side, top and bottom. Only the right and left sides needed hammerforming. The top and bottom were just bent strips. They were welded together and metal-finished for appearance. Next, I began to assemble the manifold.

I bolted base plates to a jig. The runners were tack-welded in position. Connecting pieces of the base plates were welded in. After all six runners were in place, I welded them to each other where they met at the plenum. Then I welded them completely at the base. This process strengthened and

When the plenum top is on and all welding is done, the manifold can be removed from the fixture.

The large flat surface over the plenum is for a name plate. The arrow indicates a bolt-on fuel rail near the base of the runners.

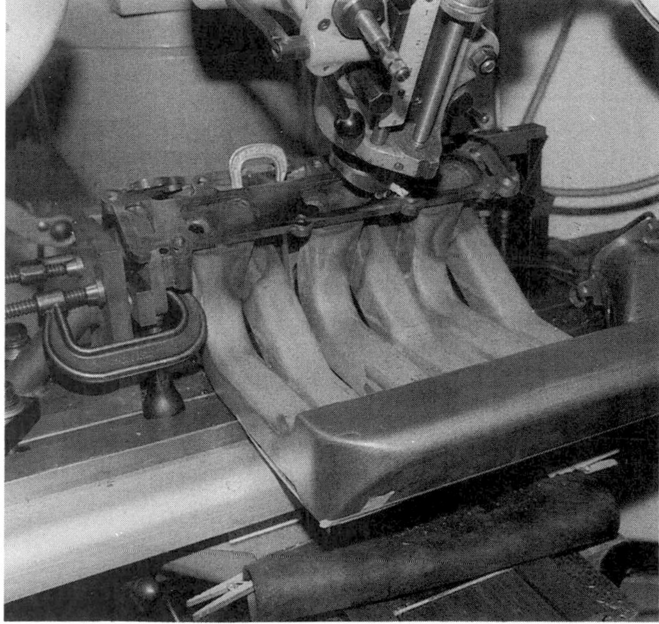

Final machining was done in a mill. All precision surfaces were cut. Other operations like spot facing, drilling and tapping were also done at this time.

The aluminum name plate, visible from the front of the finished manifold, sits atop the plenum.

unified as much of the manifold as possible at this point.

Next, I bent up and connected the plenum area. When I finished welding the plenum onto the manifold, I fabricated and welded the throttle-body neck in place. At this point the manifold went back to our machine shop for any final machining to ensure the surfaces were smoothed to the required blueprint specifications. The machinist also did all the necessary drilling and tapping to complete the manifold.

The finished manifold worked so well that we were asked to make more. Those reusable steel hammerforms got a real workout, and showed no signs of wear. The time and effort I invested in making them really paid off. Obviously, you'll need to master hammerforming techniques, as well as develop good welding skills, before you'll be able to build your manifold this way. But with a bit of practice, it is possible.

CHAPTER

RIVETING

I don't know just how far riveting goes back in history. I've seen sets of armor made in the 1500's using steel rivets to hold some parts together. I do know that today, riveting remains the most widely used mechanical fastener in prototype and manufacturing assemblies.

Riveting has come a long way since the armorers used rivets hundreds of years ago. The first rivets were made of soft steel, brass or copper. They were most likely just pins with the ends hammered down to form a crude head. They may have been crude, but they worked very well. As the years passed rivets became more widely used, particularly in the 1800s during America's Industrial Revolution. It was a strong and effective way to hold metal pieces together.

By the turn of the century, rivets were being used on everything from shoes and harnesses to bridges and ships. The use of rivets rose in proportion to the increase in manufactured goods. It was about this time when the first aluminum rivets were developed.

In the early 1900's, a major aluminum company in the United States manufactured aluminum pots and pans as its major product. These pots and pans had handles secured with aluminum rivets. Their customers soon wanted aluminum rivets for other applications. The aluminum manufacturer therefore developed many types, sizes and alloy combinations to serve the growing market. Soon after, the aviation industry was born, which had a profound influence on the development of a wide variety of rivets. Weight control and strength were key factors in early aviation. Aluminum rivets fulfilled the need.

During World War II billions of rivets were made for military aircraft. Today, there are two entire Alcoa Aluminum factories making rivets and many smaller manufacturers as well. This may give you an idea of how many rivets there are and how specialized they can be.

Advantages—Riveting offers several advantages. It is relatively inexpensive, and it also affords a method of joining different types of metal together—metals otherwise incompatible for welding. Riveting also has high *sheer strength* or high resistance to breakage. Once driven, a rivet needs no additional work. Riveting can also be done quickly. So, to sum it up, rivets are relatively inexpensive, join different metals easily and quickly, and they are strong.

TYPES OF RIVETS

A rivet has three major parts: it has a *head*—a rounded end formed during manufacture; a *stem*—a rod of metal formed during manufacture; and it has a *shop head*—formed during installation. Each kind of rivet has a difference in one or more of these elements. Rivets can have different stem diameters, head shapes, smooth or grooved stems. The rivet itself can be made from several different metal alloys.

A typical rivet catalog may list as many as seven different kinds of rivets. The two kinds of rivets I think

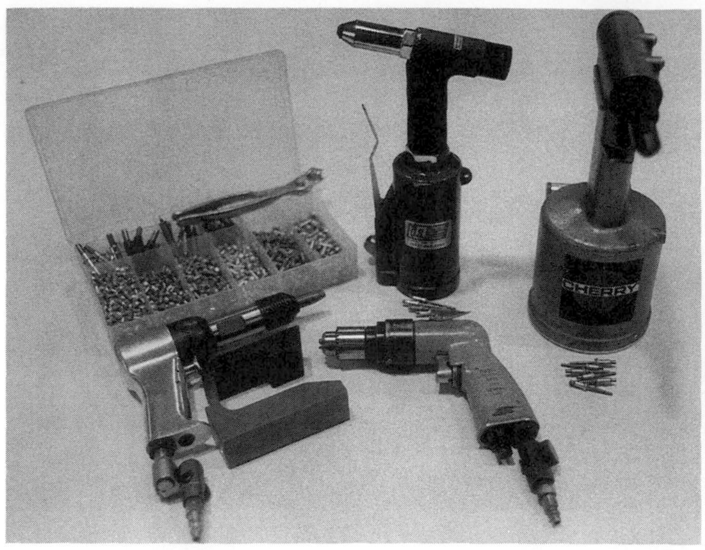

These are some of the tools used in riveting. Each has its own special application. Pneumatic rivet guns need a pressure regulator to maintain the right operating pressure.

I riveted a stiffening bracket to the side brace of this aluminum race car seat. 1/8-in. rivets were perfect for the job, spaced 1-in. apart.

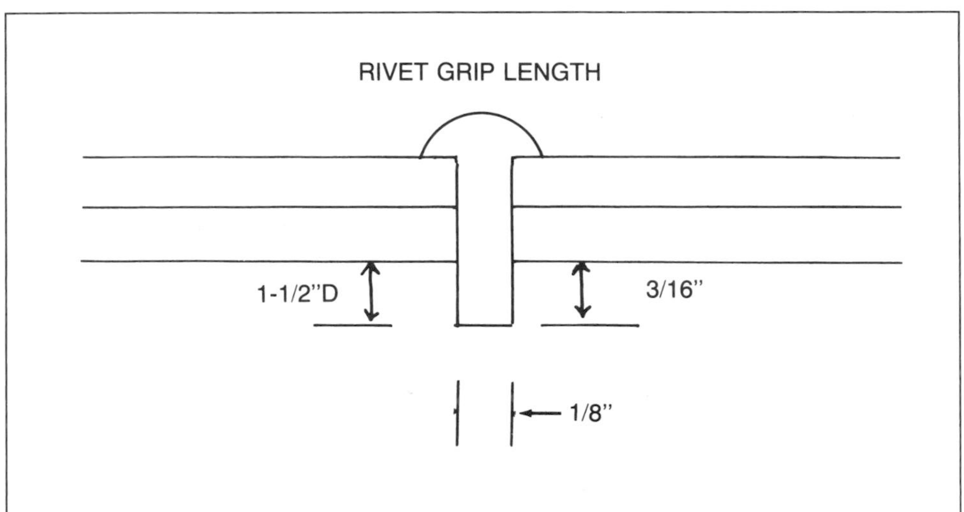

Grip length of a rivet means the total thickness of metal a rivet can join and still form a correct shop head.

A hydraulic rivet gun will pull steel rivets up to 3/16" in dia. It will save a tremendous amount of energy pulling large rivets.

you need to understand are *blind rivets* and *solid aircraft rivets*. These are the two types most often used in aircraft, race and custom cars. Each type of rivet requires special tools to install.

BLIND RIVETS

Blind rivets are so named because of the locations in which they are used. They are made to fit in situations where it isn't possible to see, or more importantly, reach both sides of the rivet. This instance happens quite a bit. The most common type of blind rivet that you are probably most familiar with is the *pop rivet*.

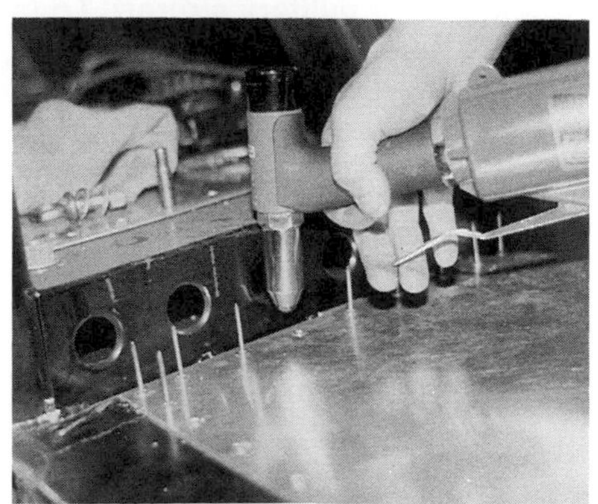

This air-powered pop riveter can pull 3/16-in. steel rivets as fast as you can pull the trigger! It's from U. S. Industrial Tool and Supply Co.

Blind rivets can be set with hand tools or with air-powered hand tools. A hydraulic hand-riveter was recently developed for blind riveting.

The choice of tool used to *set,* or install, the rivets is determined by the size, type and quantity of rivets required. Small blind rivets, such as 3/32- or 1/8-inch, are easily set with hand tools. The only real reason to get a power riveter is if you are setting many rivets daily. If you only need to do a few on a small project, the expense of a power riveter is unnecessary. Of course, it does make the job easier and saves time.

However, I strongly recommend a power riveter on rivets 3/16-inch and larger. Setting 3/16-inch rivets by hand can really be physically tiring. It

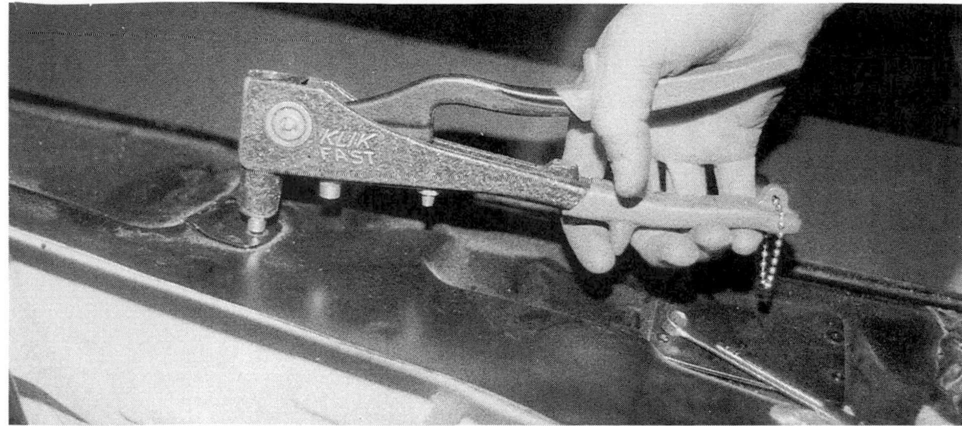

Klik-Fast hand rivet tools are very popular, for good reason. They will pull four sizes of rivets and they have long handles for good leverage.

is hard work. A power riveter able to set 3/16-inch rivets or larger ones will also work beautifully on smaller sizes of rivets. If you get the power tool it will work on most sizes.

Some aircraft blind rivets *must* be set with power tools because of the special stems which require great force to set. These stems have many grooves on them so the rivet gun can pull the rivets without slipping on the stem.

Blind rivets are often used to attach non-structural parts. They may not be as strong as solid rivets, but they more than adequately attach non-structural components firmly. They can easily endure the stresses on these parts. It is a fast, strong and reliable joining technique that is very versatile.

The Cherry Rivet company manufactures a complete line of aircraft and aerospace rivets. These rivets come in many sizes and styles. They offer optimum strength and work well in high vibration areas. Cherry also makes riveting tools for all their rivets.

SOLID RIVETS

Solid riveting is the most common method used in the aircraft industry. It is also very common in race car and other types of sheet metal work. It is basically a simple process. A rivet is placed through matched holes in two pieces of metal that are to be joined. After placing, a second head is formed on the rivet to set it. This second head—called the *shop head*—clamps the two pieces of metal tightly. Solid

SOLID RIVET SELECTION

It is important to carefully choose the rivet best suited for the particular project at hand. Choosing the right rivet results in a stronger and more professional product. Here are some general principles to guide you.

Never use a steel rivet on aluminum parts, or an aluminum rivet on steel parts. It is a cardinal rule of riveting to match the type of metal in the parts to the type of metal in the rivet.

Whenever possible, choose rivets with the same alloy number as the material you want to rivet. Say you are riveting something of soft aluminum, like 1100 series. You want to use rivets made from 1100 series aluminum. At the other extreme, if you are trying to rivet 6061-T6 aluminum, choose a rivet with comparable hardness. The idea is to make the rivet very compatible with the metal parts it's joining.

The length of the rivet is measured from under the head to the end of the stem. To determine the length of rivet you need for a given job, you add the combined thicknesses of the metal to be joined. Then you must add 1-1/2 times the diameter of the rivet stem. This allowance provides the correct amount of rivet extending through the metal pieces to form the shop head perfectly.

For an example, let's say you want to rivet two pieces of 1/16" aluminum. Adding the thickness of the two pieces equals 1/8". If you're using a rivet with a 1/8" diameter rivet stem, add 3/16" to the 1/8". This equals 5/16". You need a 5/16" long rivet to do the job.

The 2117-T aluminum alloy rivet is a good choice for general repair work. It is fairly soft and reasonably strong. It is corrosion resistant and requires no heat treatment. I choose this rivet when I don't know the alloys of something someone else originally built.

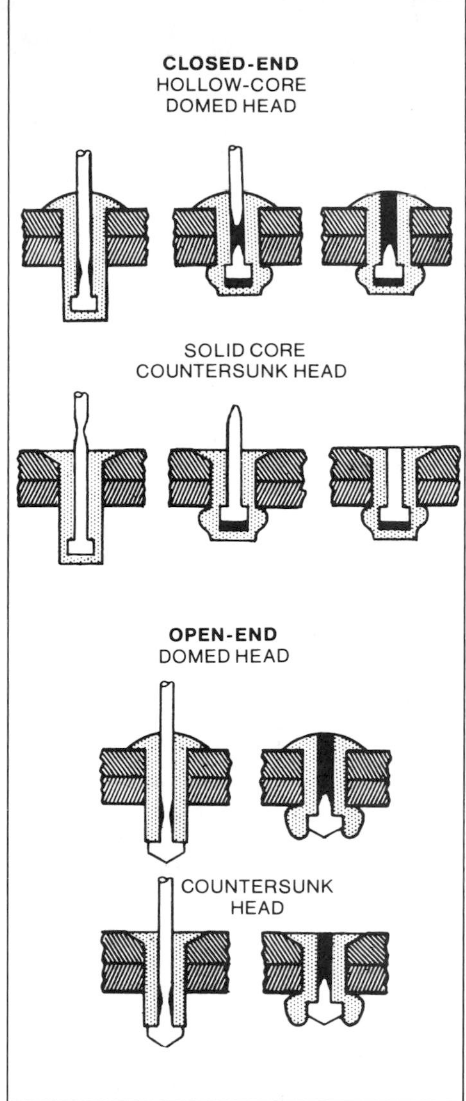

Pop rivets are hollow with a mandrel, or nail inside: Some pop rivets have closed ends and some have open ends. There are also different heads: domed and countersunk. Countersunk heads install flush with the surface. In all cases, the pop rivet gun breaks off the nail after it expands and tightens the rivet in its hole.

rivets are most commonly set with a pneumatic rivet gun and a *buck*. The rivet gun also has a part called a *set*—a part fitting onto the rivet head and transferring power from the rivet tool to the head. The *buck* or *bucking bar*—a solid tool to back up the rivet—is held against the rivet stem. The force of the gun and the weight of the buck form the shop head on the rivet as it's set into the holes in the metal pieces. The rivet expands to lock the joined parts tightly. The solid rivet is much stronger than a pop rivet for these reasons.

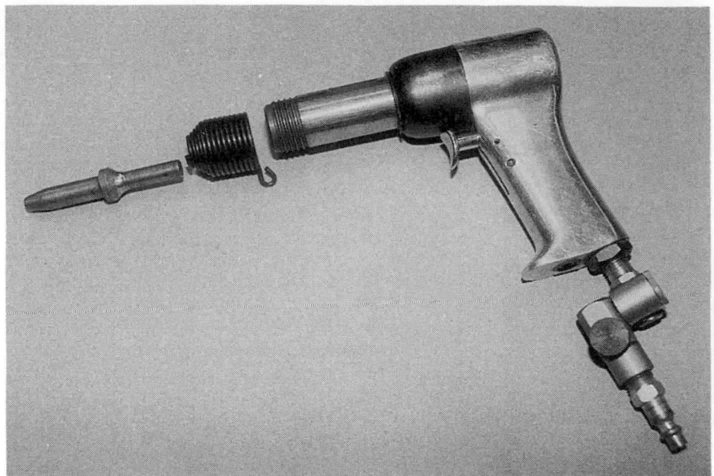

A U. S. Industrial Tool and Supply Co. pneumatic rivet gun model TP83 is a very popular size. It works conveniently for home aircraft builders.

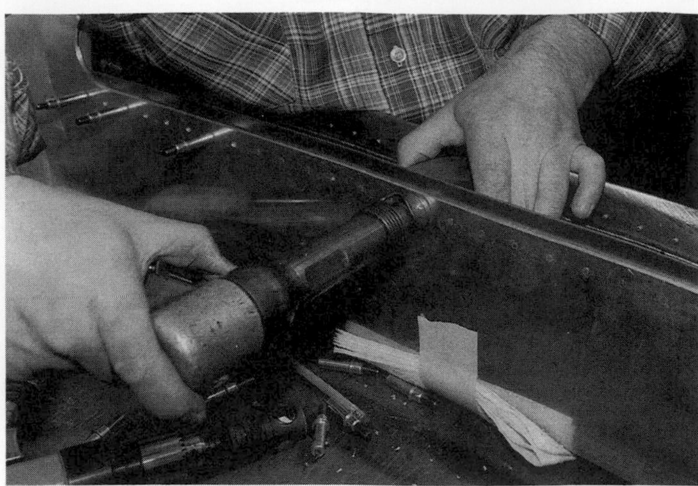

I'm tightly holding a bucking bar directly behind the rivet gun, exactly on center. This ensures an even shop head on the rivet.

PNEUMATIC RIVET GUNS

Pneumatic rivet guns are available in various sizes and shapes. The rivet gun is composed of three elements; the gun itself, the *rivet set*—a special tool fitting into the gun barrel—and the *retaining spring*. The *capacity* of each gun, which determines the maximum size rivet that can be set, is usually stamped on the barrel. Pneumatic rivet guns operate on compressed air pressure from 80 to 100 psi. When you buy a rivet gun it is likely it will come with several *rivet sets*. Rivet sets are special hardened steel tools fitting into the rivet gun barrel. The rivet set has a head formed to the exact shape of a particular rivet type.

For example, three common rivet sets are: *universal, round* and *flat*. Each one fits a particular shape of rivet head. There are many other kinds of rivet sets, each intended for use with a special kind of rivet head. *It is important to match the rivet set to the proper, corresponding rivet*. Otherwise, the rivet head will be deformed as it is installed, which not only looks poor, but compromises strength.

Keep the rivet sets clean and dry. The part of the rivet set fitting over the rivet head is a very precise, polished shape. Dirt or rust on this surface will mar the rivet head. I keep lightweight oil on my rivet sets and clean them before and after use.

A rivet set is fairly easy to identify. Some have code numbers stamped on them, which refer to the type of rivet head they match. Others may actually spell out the type and size of rivet head they can handle. The stamped words might say "1/8-inch universal," for example. The manufacturer of the rivet set provides a sheet explaining the code numbers, if the set is stamped with a code rather than words and measurements.

No pneumatic rivet gun should be without a swivel type air regulator. It makes it easier to use the tool because it swivels the air line out of your way. Photo courtesy U. S. Industrial Tool and Supply Co.

The rivet set is held into the gun with a special spring which screws onto the barrel. This spring is known either as a *beehive spring* because of its shape, or as a *retaining spring* because of its function; depending on how you look at it, of course.

RIVET BUCKS

A *rivet buck* is the piece of hardened metal placed against the rivet stem to form the shop head. Sometimes it is called a *bucking bar*. Choosing the correct buck is very important, as it can deform the shop head if it doesn't have the right shape. If a buck is too light the rivet may sink below the metal surface as it is being set.

A buck can weigh as little as a few ounces or as much as ten pounds. The weight depends on the size rivet to be set. There is no magical chart to guide what weight buck to use for what size rivet, but I can offer a few general guidelines. Quite often, trial-and-error is the only method to determine the correct rivet buck size.

The most common solid rivet I use is a 1/8-inch aluminum. For this size, I use a two-pound buck. For a 1/8-inch *steel* rivet, I need a much heavier buck, about four pounds. The general rule is to use a buck twice as heavy for steel as for aluminum. I know a buck is satisfactory when the rivet gun sets the rivet quickly without prolonged hammering. If it is going and going and the shop head isn't forming, it is time to use a heavier buck.

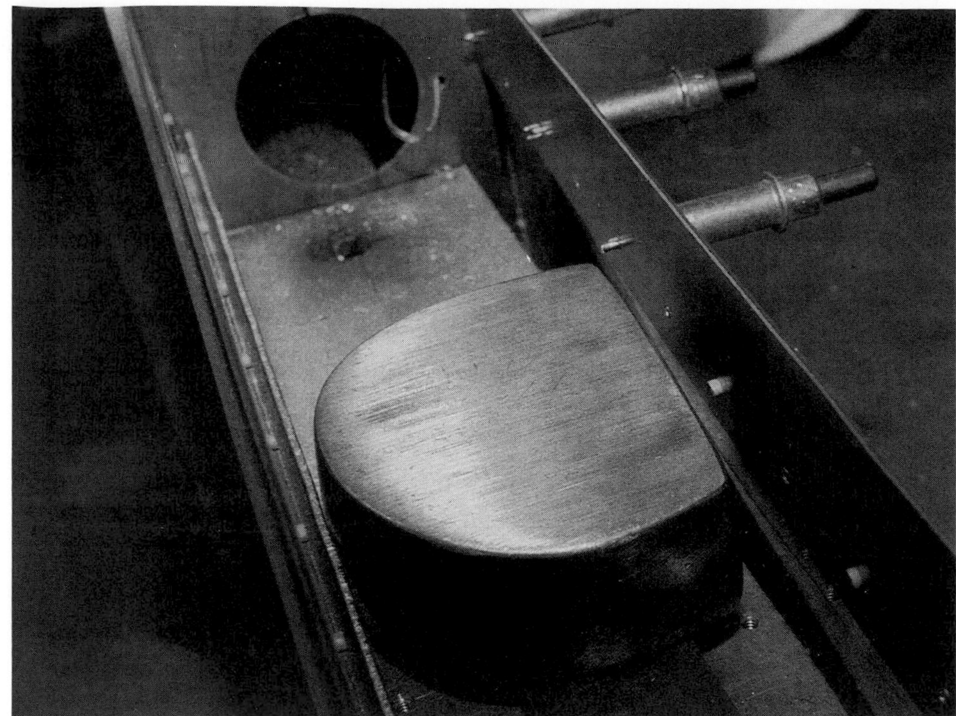

A heel dolly was perfect to buck these rivets. The weight was right, and the size and shape fit into a close area. Many bucking bars would have been too long to fit.

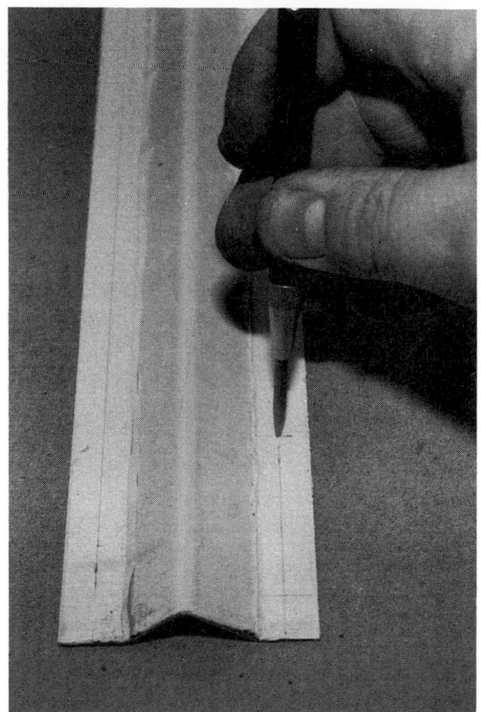

Use a center punch to mark the position of each rivet. Randomly placed rivets look sloppy.

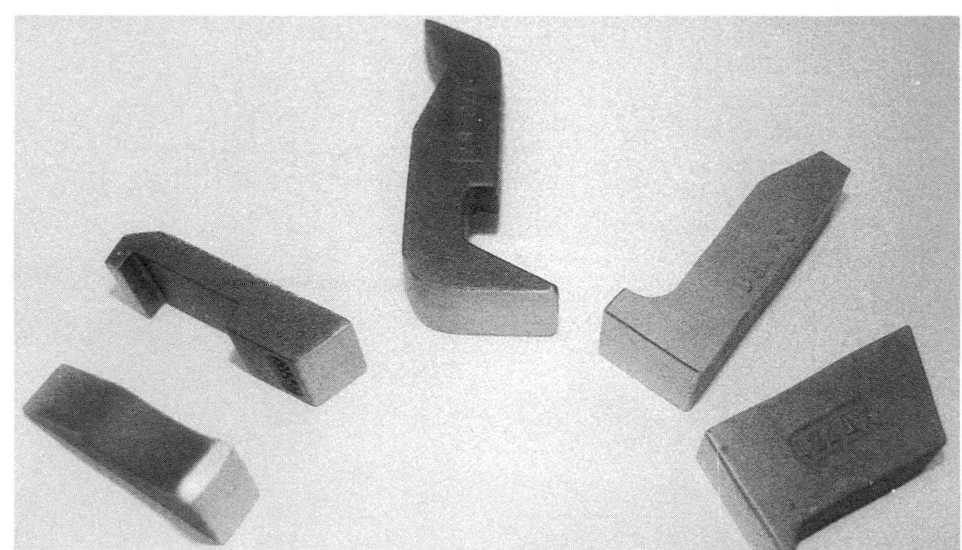

Sometimes you need a special shape bucking bar to back up a rivet in an awkward place. That's why a wide assortment of bucking bars are available.

Using The Buck—Hold the buck tightly against the rivet stem at a 90-degree angle to it, or the stem will bend instead of forming a shop head. The buck has to be held tightly until the rivet is completely driven. If you move or release pressure on the buck too soon you'll sink the rivet. Any rivet you sink below the metal surface has to be removed and replaced.

Various aircraft tool suppliers offer a wide range of bucks, or bucking bars. They come in dozens of sizes, weights and shapes. None are very expensive, and most are reasonably priced. There is nothing wrong with using a dolly in place of a buck. Dollies are okay if the weight and shape fit the job to be done.

You can also make your own bucks. You may have a special requirement for space or weight. Just be sure the homemade buck is smooth and scratch-free with no sharp edges. Make it from round or square steel stock, depending on the shape you need. The steel doesn't need to be heat-treated or hardened. Mild steel will hold up fine.

PREPARING RIVET HOLES

Rivet holes must be the right size. If the hole is too large the rivet will not completely fill the hole, and the joint will not have its full strength when the rivet is set. If the hole is too small the protective coating on the rivet can be scraped off as it is pushed in. You'll also waste alot of time fiddling with the rivet trying to get it into the hole.

RIVETING

Adjust the speed of the rivet gun by pressing it firmly against a wooden block, without rivets, before you begin to rivet. This way you set the speed without wasting a rivet head. *Remember never to operate the gun without resistance against the rivet set.*

Set up a trial rivet situation. Take two scrap pieces of the same metal as

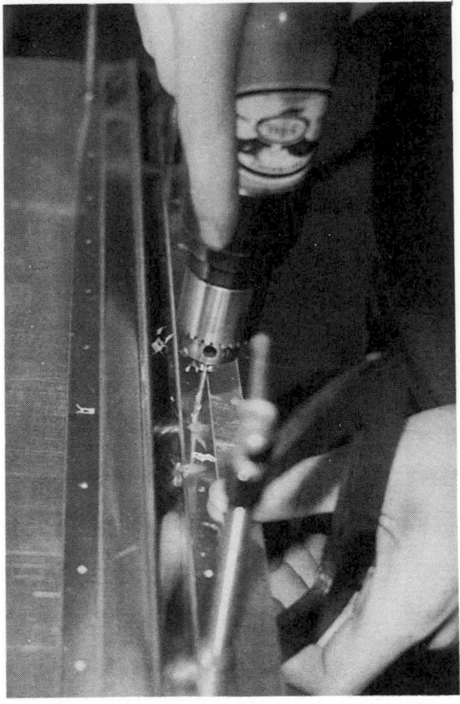

Clamp the metal parts and pre-drill the rivet holes.

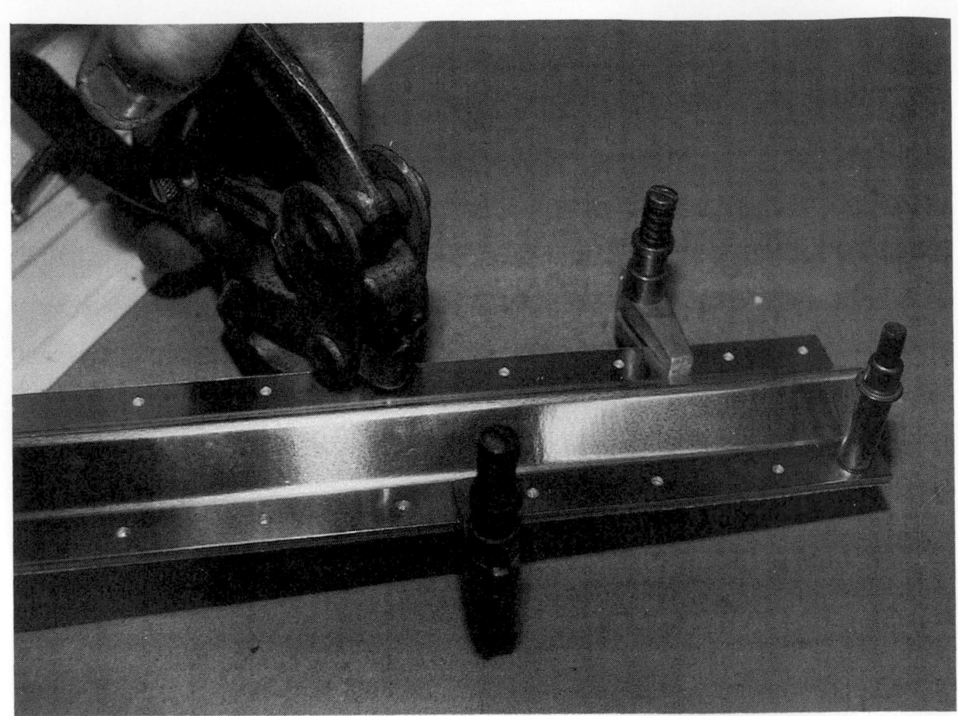

After pre-drilling you can either final drill the rivet hole to the rivet diameter, or use a hand lever punch the same size as the actual rivet diameter. Punches leave a burr-free hole.

the project. Drill rivet holes as I described. Adjust the rivet gun speed against a wooden block. Take some rivets of the type to be used, insert each in a rivet hole. Use the buck selected, and set the rivet.

This process will show you if all the variables are right. It will show whether the holes are the right size, and whether the speed is just right. It will determine whether the air pressure for the gun is right or if the buck is heavy enough. Finally, you'll know if the rivet is the right length. This is the time and place to make mistakes and correct them without risking the whole project over a small error.

When you are satisfied the trial set up is working just right, go to the project itself. Repeat the successful pattern of rivet holes, gun speed, rivets and buck. Use plenty of *Clecos*. They help keep the parts held tightly together during riveting. The last thing you want is to have the pieces slip or separate during riveting.

MAKING RIVET HOLES

To make the right size rivet holes follow these rules:

1. First mark the location of the rivet holes with a center punch big enough to keep the drill from wandering.
2. Then drill a hole about 1/32-inch *smaller* in diameter than the intended rivet diameter. This procedure is called *pre-drilling*. The undersized hole is called a *pilot hole*—it guides the drilling of the final hole. Drill the pilot hole carefully.
3. The next step is to drill the pilot hole to the correct diameter for the rivet. Be careful not to let the hole get too big by letting the drill bit wander.
4. The final step is to deburr the final hole carefully. Remove any rough spots or irregularities *without enlarging the hole*. If a rivet is set against a burr, the burr can make the rivet work loose later on.

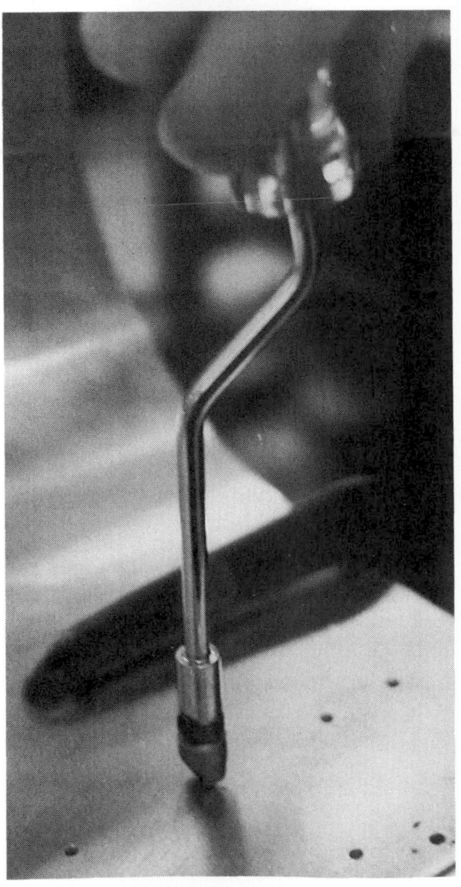

Deburr—remove any rough surfaces from—rivet holes. It ensures a tight fit.

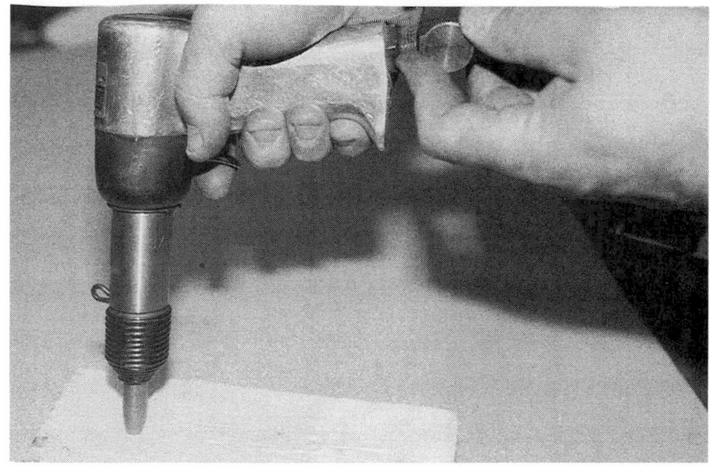

Set the rivet gun pressure safely by pressing the set against a piece of soft wood.

Don't rivet the *first* time on your project if you're a novice riveter. Do a practice rivet series on two metal strips like the metal in your project. Make adjustments as necessary.

MORE ON RIVETING

It is beyond the scope of this book to cover the complex rules and guidelines about riveting in their entirety. There are numerous government guides concerning aircraft riveting. There are many guides published by the rivet manufacturers, the Air Force and the Navy. The Air Force and Navy ones are training manuals. I recommend them highly if you can find them. They deal mainly with air frame repair, but the principles and specifications are applicable to other kinds of metal work. I have found them in used book stores and sometimes at used tool shops or flea markets.

Another handbook I value is the *Aircraft Mechanic's Shop Manual*, edited and compiled by Larry Reithmaier. He used numerous government and manufacturer sources to write the book. Some tool suppliers carry this book, including U. S. Industrial Tool and Supply Co. (see address in the *Supplier's Index)*.

I have also obtained information about riveting from the Experimental Aircraft Association. Their publication *Aircraft Sheet Metal* is available by writing to: Experimental Aircraft Association, Catalog Sales, Whittman Airfield, Oshkosh, Wisconsin, 54903-3086. You'll find it especially useful if you work on aircraft sheet metal, although much of the information also applies to automotive work.

You might have noticed by now that I didn't cover the various methods of welding automotive sheet metal in this book. If you'd like to learn these techniques, I recommend my first book, the *Metal Fabricator's Handbook*, which is also published by HPBooks. I cover both gas and electric welding on aluminum and steel. HPBooks also publishes the *Welder's Handbook*, a complete guide to all welding techniques. Both are available at your local bookstore, or by writing to Price Stern Sloan, ATTN: Customer Service, 360 N. La Cienega Blvd., Los Angeles, CA 90048. Outside California order toll-free (800) 421-0892.

As you rivet, stop and check the shop head. It may need another burst from the rivet gun to make it identical to other rivets. Check as you go.

CAUTION!

Never run the rivet gun without placing it against something. If you use the rivet gun without a backing, the vibration may break the retaining spring. When the retaining spring breaks the rivet set can fly out of the rivet gun and could very possibly hurt someone. Use the tool right and avoid any chance of an injury. If you need to test the gun, place the rivet set against a block of soft wood. Then you can test it safely.

CHAPTER

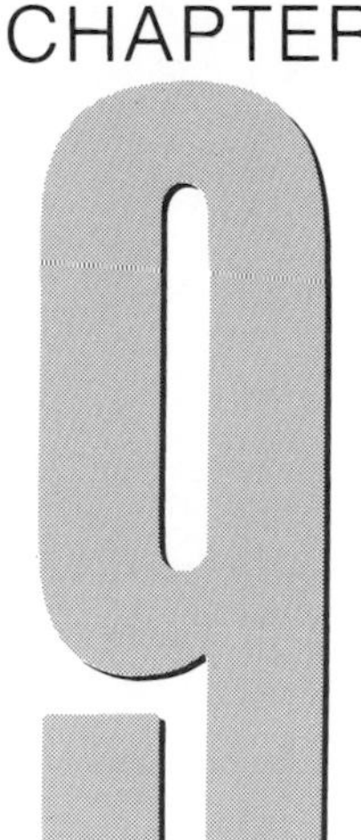

RESTORATION

There are three basic reasons why people restore cars. The first is simply *nostalgia.* Today's baby boom generation is now in a financial position to finally afford the "dream" car they always wished for back in high school: be it that '57 Chevy Dad let them borrow for dates; or the '32 Ford Coupe the "coolest guy" cruised downtown in Saturday nights, classic car restoration is often-times a personal decision.

The second reason restoration and preservation of vintage vehicles is close to some people's hearts is sheer *aesthetic enjoyment.* Many car buffs sincerely believe the quality and beauty of older vehicles are reason enough to justify caring for and restoring older cars and trucks. I often agree with their judgment.

The third reason is *financial.* Collector cars represent one of the safest, and most profitable investments on the market today. For example, a pristine '57 Chevy convertible is worth $40,000 today. In the year 2000, it is estimated that the worth will be well over $100,000. The same can be said for a '67 L-88 Corvette, or a '67 Z28 Camaro. A Ferrari GTO could be bought just 10 years ago for $15,000. Last year, one sold for $1.5 million. A Ford 427 Cobra cost just $6,000 dollars on the showroom floor back in the Sixties. Today, you'd have to part with $1 million to park it in your garage. As of yet, no CD, IRA or Mutual Fund has returned that much on an investment in such a short period of time. Besides, you can't drive and enjoy a Mutual Fund. In all seriousness, these are very valid reasons why more people are turning to classic car restoration than ever before.

The value of these cars is often proportional to how much of the original car is intact—which includes the sheet metal. "Restoration" is defined as; *the act of putting something back into a prior or original position, place or condition.* The definition is simple and straightforward enough, but the key word is *original.*

In order to truly restore a classic car, it is essential to pay attention to details and to make sure that every part of the car is correct, just the way it rolled off the showroom floor. Unfortunately, many people just "patch and fill" portions of the car and call it a restoration. This work may suffice for amateurs, but it isn't original.

What I'd like to do is offer some advice that should help you out with any restoration project.

PATTERNS AND INFORMATION

Patterns and information for restoration should be 100-percent accurate. If a part is missing, first look at the opposite side of the car. See if you can get enough information from the opposite side to make the mirror image of the part. Most body parts come in right and left. Take advantage of this. The opposite part can supply you with all the correct information you need to replace a missing piece.

REVERSE PATTERNS

Making a reverse pattern of a part is easy. For example, the bends in a

The restoration of this Austin Healy included repairing extensive rust damage to the body panels.

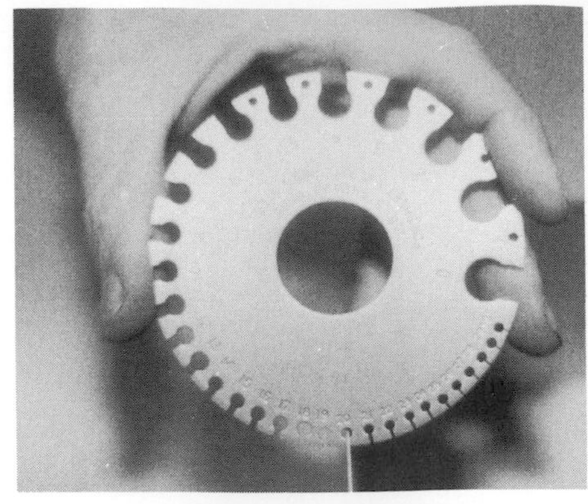

Another way to check metal thickness is to use a sheet metal gage. Eastwood Co. sells this handy tool.

splash apron all go in one direction. If there is one good splash apron on the car, and the other splash apron is missing or damaged, it is possible to make a pattern for the replacement apron from the good one. Once the pattern is made, take the pattern, transfer it to metal, and bend the metal *exactly the reverse* from the one you measured, to form an opposite side. This process sounds alot more difficult than it is.

Sometimes, for one reason or another, both sides of a given part may be gone from the car you're restoring. When this happens, you need to do a little footwork.

REPRODUCTIONS

Of course, the more original sheet metal that can be repaired, retained and restored, the better. However, sometimes this just isn't possible. There is now a multi-million dollar industry that involves the reproduction of parts for classic cars that are no longer available from manufacturers —including sheet metal. Some parts are easier to find than others, depending on the popularity of the car. Start out by checking all the manufacturers of reproduction antique auto sheet metal parts. If the car is a common one, then you should have no problem finding a reproduced part, such as: fenders, floorpans, firewalls, wheelhouses, splashpans, runningboards and many kinds of lower fender patch panels. Purchasing these reproduced parts can save time and money.

If you happen to be restoring a car like a 1937 Pontiac it is not likely

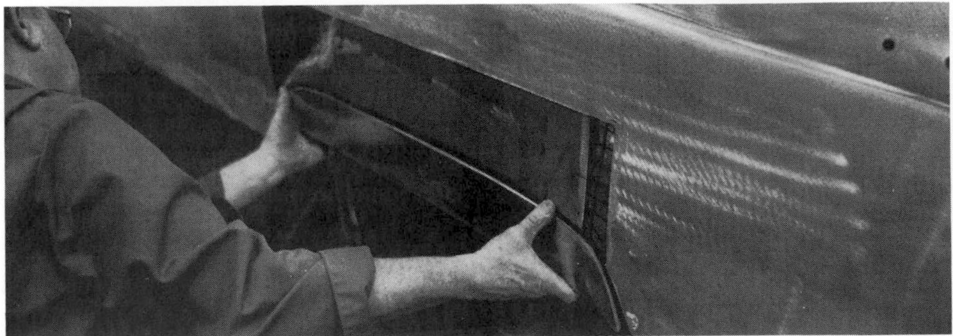

Patching a fender panel at Hill and Vaughn's restoration shop in Santa Monica, California is meticulous work. This craftsman is checking to see how the patched area will match the upper curve of the fender skirt.

KEEPING IT AUTHENTIC

You'll want to keep your restoration as authentic as possible. Here are some often over-looked points that can make the difference between 1st and 2nd place in a top car show.

1. Make your sheet metal restoration work as authentic as possible. Study all the aspects of the project, and obtain as much manufacturing history, facts and figures on the car as possible. Car clubs, enthusiast magazines and libraries are good sources.

2. Avoid the use of plastic fillers on pre-'55 cars. Plastic fillers weren't around until 1955.

3. Don't use pop rivets for the same reason. Pop rivets weren't commonly used on cars until the early '60's. Although some cars, like the Cobra, use them, many others do not. Research your particular vehicle and decide whether or not pop rivets were originally used when it was new. Many of the cars prior to pop rivets used solid rivets. Solid rivets are not hard to get or install. Use solid rivets on any car which would have had them when manufactured.

4. Don't use a different type of welding when restoring a car than was originally used. No 1935 Packard had the floorpan welded in with a Miller Heli/arc. It was welded with oxyacetylene, which should be used to restore that car.

5. If you use incorrect rivets, welding or filler on a restoration intended to duplicate original condition, an expert—like the judge in a car show—will spot the irregularity. You'll lose points. And, on a more personal level, if you make mistakes riveting or welding or filling, you'll know you have compromised the quality of your restoration. Quality is the final reward of careful work.

you'll be able to find reproduction parts for it. It is an uncommon vehicle for restoration. That doesn't mean it isn't worth restoring. It just means there aren't as many of that particular vehicle around, or they appeal to only a few people. Manufacturers generally do not concentrate on reproducing parts for those vehicles.

RESTORATION SHOPS

There are a number of professional restoration shops all over the country. They are generally staffed by real metal craftsmen. However, their services do not come cheaply. A true metal restoration shop will not only be able to repair body panels, but will also have the capability to build them from scratch. All the same tools and equipment found in a large sheet metal shaping shop, such as English wheels, power hammers and Eckold Kraftformers, should be in a restoration shop. If they don't have this equipment, then they may not be able to produce the panels you need.

Sometimes a restoration shop will agree to make new panels for an antique car, but promptly sub-contract the actual fabrication to the local metal shaping shop. Of course, that means you are paying a profit margin to *two shops*. Make sure the shop you approach is capable of performing the work in-house.

RESEARCH

Sometimes, when a car is a rare year and model, a large portion of the time spent on restoration is research time. Keep in mind your final goal is to restore the car to its *original condition*. Some places to get information on your car include car clubs, private collectors who own a similar car, auto museums, auto books and auto magazines. Most of these books contain excellent photos and accurate information. They also include line drawings showing accurate details on specific models. Most car enthusiasts are only too willing to share information, or refer you to others who may have information to help you.

RESTORING SHEET METAL

Let's assume that you've gathered all information regarding the car, located and purchased the reproduced parts that couldn't be repaired or were missing, and are now ready to begin returning the car to "better-than-new" condition. Though the project ahead is no easy task, these following procedures should help the project go smoother, and help avoid some costly mistakes.

USE THE SAME METAL

The metal chosen to repair or fabricate parts should be as close as possible to the original metal used when the car was manufactured. The metal thickness can be determined by measuring carefully.

Choose a flat section with an open edge that is free of rust, paint and grind marks. This type of section will ensure an accurate micrometer reading. Place the micrometer over the spot and read it. Then move around and read a few more similar places.

If the readings are all the same, convert the measurement to a standard metal gage size by cross-referencing it with a *sheet metal gage chart*. If the readings vary, examine the surface carefully to see if you have taken some readings over a burr or a thin spot. Check again for consistent measurements with the micrometer.

For instance, let's say the micrometer reading was a consistent 0.032-inch. The chart listed in the back of the book, lists the common gages and fractional equivalents. It says this is equivalent to 21-gage.

Knowing the gage is important when ordering metal. If you ask for metal in thousandths, the salesperson may try to sell you metal that is close but not exactly what you need. Be specific and firm about the gage. If they don't have it in stock, ask them to order it, or try another supplier. Do *not* settle for something that is "close" to what you need.

CLEAN WORK AREAS

Always be sure to clean the area you are working on. The area must be clean in order to do sound welding. *Never weld in an area that has paint, rust or any kind of dirt or contaminant on it.* Welding on dirty metal makes the welding more difficult. The welds will look poor and be pitted. As you weld the torch will pop and may blow holes in the weld. Your welds won't look nice and will be harder to metal-finish.

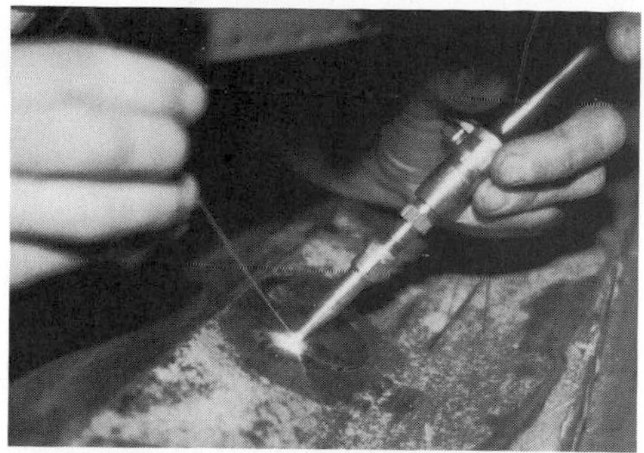

A Dillon Mark IV welding torch makes it easier to weld thin sheet metal like this fender patch because it has a fine, pin-point flame. It's also very good for gas welding aluminum.

Smoke and the danger of fire constitute another real problem. Welding on dirty metal will make the paint burn. The smoke may be toxic, may burn your eyes and make it hard to breathe. The heat from a welding torch can also melt undercoating, which will then drip on the floor. Undercoating burns easily and could start a nasty fire. *Always weld in a safe, well-ventilated area with a fully-charged fire extinguisher close by.*

FITTING, TACKING AND SMALL WELDS

Take time to fit the parts correctly before tack-welding them. Begin by aligning the part so all the edges are in exact position. Check it thoroughly to make sure it fits perfectly.

Then use plenty of vice grips or other firmly-secured clamps to hold the part in place. When the part is exactly where it needs to be, begin tack-welding.

The first tack-weld should be *small*. Before making the second tack, check to see if the part has shifted or pulled out of position. If it has moved even the smallest amount, reposition it and start over. If it did not move make your second tack. Check again for any movement. Keep up this process of tacking and checking until the part is

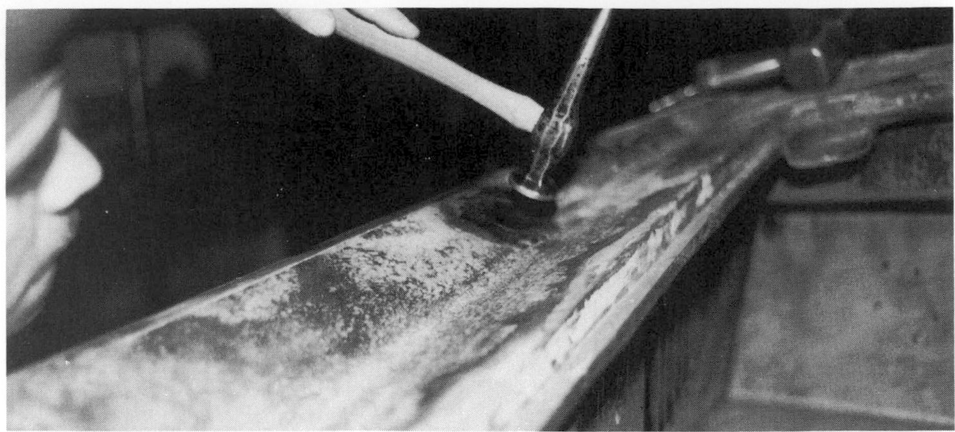

Don't stretch the metal in a patch by hammering too hard. Use the hammer, and the dolly held underneath, to level the patch's surface so it matches the fender.

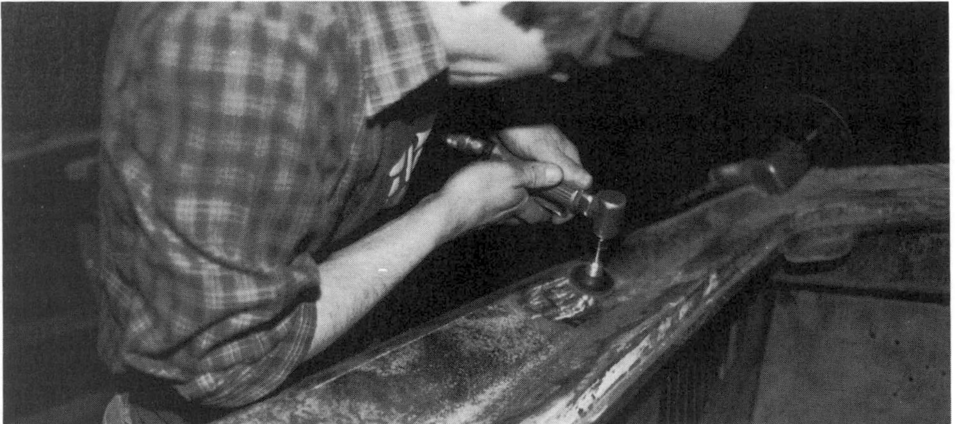

Be careful removing excess weld from a patch. Do *not* remove any of the metal surface. After initial grinding you may need to use more dollying to raise low areas.

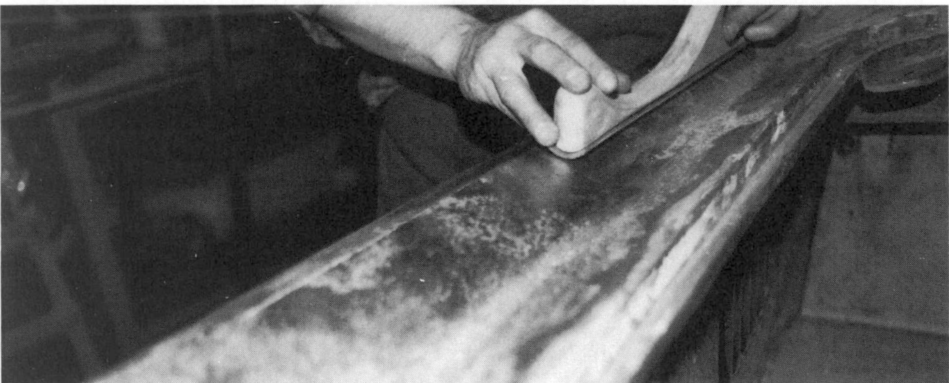

The last step in metal finishing a patch is using a vixen file to smooth the surface.

firmly positioned by many tacks.

Use plenty of *small* tacks, no more than one inch apart. Use no welding rod, or as little welding rod as possible, on the tacks. You won't need much rod if the fit is good in the first place. That's one reason a perfect fit is important before beginning to weld.

A good fit is important for another reason. The part will pull or warp when it is welded if there are gaps between the part and the metal surface. This is caused by the excess welding heat generated when welding up a gap.

A part may warp a bit after it is tacked in. When this happens, take a hammer and dolly and smooth out the area. Remember to do this after it is tacked. Final welding can take place when it is dollied back to its original shape. Use very fine welding rod, keeping in mind that the more rod you use, the more you will have to remove in order to smooth the surface. The idea is to join the seams without adding any more rod than necessary.

METAL FINISHING

After the metal has been welded in, it must be finished. No matter how skilled a welder you are, the part and surrounding area will be affected by the weld and the heat. This is where metal-finishing comes into play.

The method you choose to finish the metal is determined by how distorted the area is. Let's assume the worst: a medium-sized patch was welded in that was mainly flat on one side, but the other side was curved.

The curved part of the patch stayed fairly close in shape when the welding was done. This is because the curve in the car and the curve on the patch made it stiff and minimized warping. The rest of the patch, almost flat and joining a similar surface, did not stiffen but absorbed the warping from the whole patch. It distorted the surface, and became wavy and ugly. In the metalworking trade we refer to this as a piece of metal "going nuts". It looks pretty strange.

Dollying—This wavy, distorted area must be metal-finished carefully and in a specific order of operations. Care must be taken not to stretch the metal. In a severe case like the one I am describing, the first step is to get alot of light on the part you're metal-finishing. The light shows up all the irregularities. It is difficult to work if you can't see the area very clearly.

Use a hammer and dolly to *rough* the area back to its desired shape. Choose a flat-faced hammer and a dolly with a shape matching the underside of the area you are working. Hold the dolly in one hand and move it *under the high spots*.

Use plenty of *light* blows. The idea is to hammer hard enough to move the high places down to the dolly, but not hard enough to stretch the metal. Many easy blows are better than a few hard ones. It is skill rather than brute force that does the trick.

You should never hit the metal hard

A view of the finished patched fender (arrow) shows how proper metal restoration produces a surface needing no body filler.

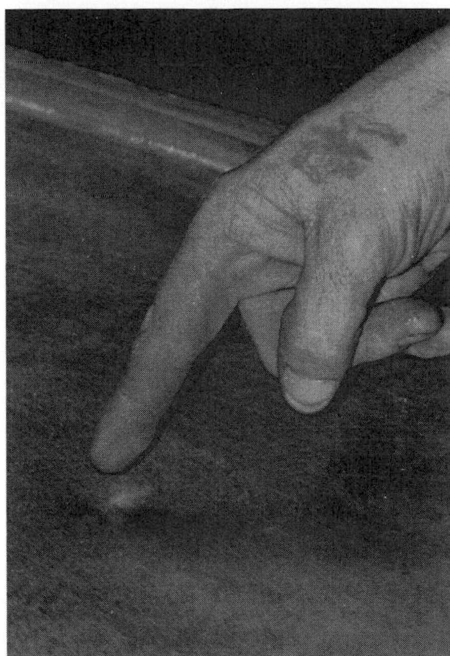

The dent in the rear panel of a '34 Packard was about 1/2-in. in dia. and 1/4-in. deep.

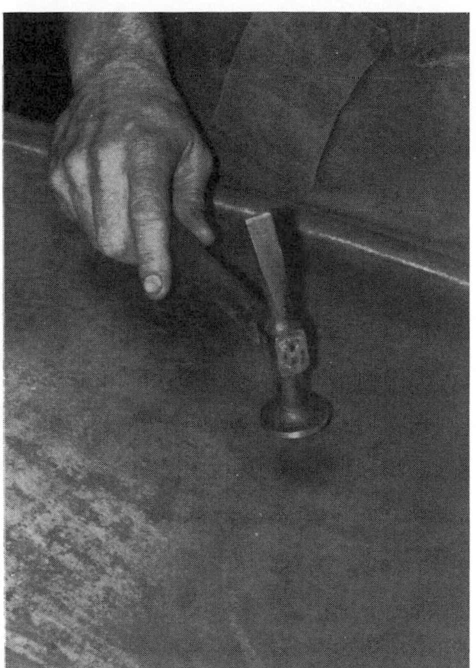

Scott Knight bumped up the dent with a dolly. Then he hammered and dollied the metal until it formed a low, smooth swell.

enough to hear the hammer ring on the dolly. You may dolly the metal more than once before you are happy with the general shape. That's fine. Stick with it. *Study the surface as you work to see how the distortion is smoothing out. Move around and look at the whole area from several directions.* The light will show irregularities in one spot or another.

Ask yourself some questions as you work. Is the curve flowing smoothly up or down in the direction it is supposed to? Is there a smooth transition between the curve and the flat area? Is the flat part of the patch truly flat where it is supposed to be?

Look the area over carefully. It should have a very smooth, flowing, shape and appearance. If not, you need to keep working until you see it.

Weld Finishing—When you are finished with the first dollying, check out the weld around the part. If it was done correctly it will be sound, even and low, without much welding rod deposit. If the weld is *high*, you must make it even with the surface. There are several ways to remove excess weld. One common way to remove excess weld is to file it off.

The problem with filing is the amount of time and effort it takes to get

Scott Knight's special shrinking and leveling disk is not a magic wand, but he thinks it's as important to a body man's bag of tricks as his hammer and dolly. Notice how a shot bag is holding the panel down while Scott works.

Scott floated the grooved area of the disk over the stretched metal. He used moderate pressure until the area began to grow outward. He kept applying pressure until the area was hot enough to make a wet rag sizzle when it was applied to the area to cool it.

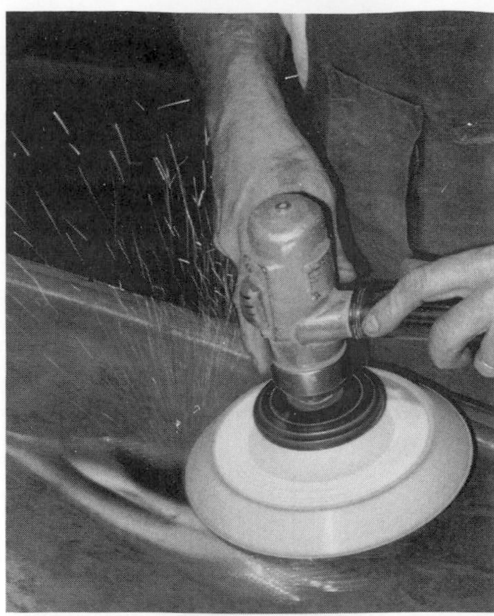

40-grit sandpaper was mounted on a soft pad sander to finish smoothing the metal.

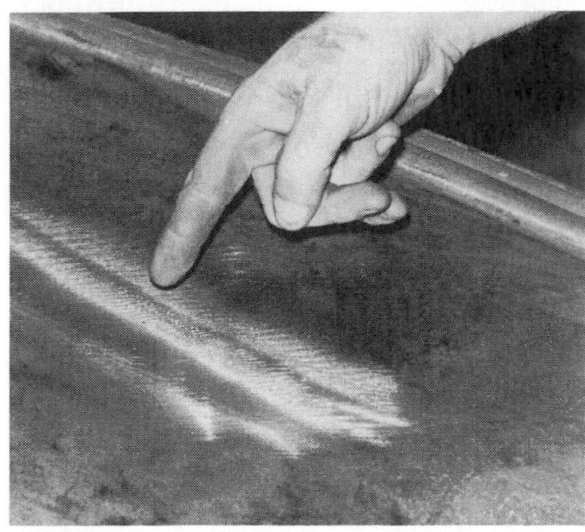

The finished repair shows the dent was completely gone, there was a minimum of metal removed, and no filler was needed.

These blanks are ready for bending. They will become replacement splash aprons—the metal panel between the bottom of the door and top of the running boards—for an antique car. Fay Butler will do this restoration. Photo courtesy Fay Butler.

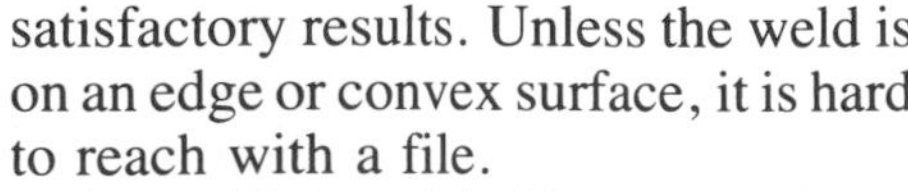

satisfactory results. Unless the weld is on an edge or convex surface, it is hard to reach with a file.

A small, hand-held pneumatic or electric grinder or sander is a better way to remove excess weld. They are fast and efficient. When you remove weld with a file or a power tool be very careful *not* to remove any metal below the original metal surface. You sure don't want to file a ditch into the metal. Making a low spot is a big mistake that will be difficult to correct.

Final Planishing—After you have removed all the excess weld from both top and bottom, you can do more dollying. This time use a small planishing hammer and the same dolly. Remember the dolly must match the shape of the metal as closely as possible. This time around, with the excess weld gone, spend time gently planishing the area. This is the second step to proper metal-finishing.

Go over the area as much as you need to. Just be sure to avoid hitting the metal too hard. If you hit it too hard you'll stretch the metal and ruin the shape you intended. Another problem with hitting the metal too hard is that it may *work-harden*, which would make it *much* more difficult to get the correct shape needed to fix the spot.

As you planish the area, run your hand across it in different directions. Your hand will feel problems right away. Occasionally you may feel a low spot that needs hitting from underneath. I rely on touching the metal

The large radius bend is being made on the splash apron. Notice Fay Butler has already formed beads in the parts. Photo courtesy Fay Butler.

Fay Butler's finished splash aprons look beautiful and are exact duplicates of the original parts. Photo courtesy Fay Butler.

Fay Butler and his assistant are using a Yoder power hammer to form a tail panel for a Pierce Arrow. Photo courtesy Fay Butler.

surface so much during metal-finishing that some guys have made jokes saying I'm massaging the car.

Body File And Sander—When you have finished the second step of dollying, use a body file and sander. Body files are also called *vixen* files, and come in many sizes and shapes.

The one I use most often is flat—about 1-1/2-inches wide and 12-inches long. Vixen files must be used with a file handle or holder. Some file holders are made of wood and some are cast aluminum. The cast aluminum ones are nice because they have a turn buckle adjuster. You can curve the file to a convex or concave shape. This comes in handy because alot of the metal you need to file is not flat, but curves up or down instead.

Use long straight strokes *in a forward direction only*, when you file with a vixen file. The file cuts smoother if it is held at a 30-degree angle. File forward and lift on the back stroke, because the file teeth face forward. Using the vixen file will quickly illustrate the high and low spots.

After a few strokes across the area, put the file down and gently work out a high spot with a hammer and dolly. The operative word here is *gently*.

This process of using the file to read the surface is repeated as much as necessary to completely work the area smooth. I put *layout dye* or *Dykem*, on the surface when I use a vixen file to metal-finish an area. I don't put dye on until I am well into the work. At first your hands and eyes pick up the high and low spots. But as you go on, those smaller irregularities get harder and harder to find. The layout dye makes it easier to see the irregularities.

The file will take off the dye very quickly. When the file hits a high spot, it removes the dye. On the same file stroke any low spots will show up because the dye won't be removed where the metal is low. This is a quick method of identifying the highs and lows, which helps to speed up the metal-finishing process.

Be careful not to remove too much metal when working on the high spots. The idea is to *move* all the uneven surfaces back to their correct, original position. A danger here is the tempta-

The finished tail panel is made of 1/8-in. thick aluminum. It duplicated the original thickness of metal in the Pierce Arrow. Photo courtesy Fay Butler.

This beautiful quarter-panel was made by Marcel DeLey. He used an English wheel to duplicate the original large low-crown surface. He beaded the panel after shaping.

tion to simply grind or file off the high spots. If you grind them off, the metal will become thin in those areas. It can thin even to the point of breaking through. Don't let this happen.

Work the metal up and down with tools until the surface is smooth and uniform. Sometimes when metal-finishing steel you will come across a high spot which will not work out by hammering and dollying, no matter how much you try. Most likely the metal is stretched. In other words, there is too much metal for a given area. This can be a real problem. You can shrink the high spot in the steel with heat.

Shrinking Steel With Heat—is done by taking an oxyacetylene torch and heating up the high spot. Hold the torch directly over the bump to be shrunk. Heat up a spot about 3/8- to 1/2-inch in diameter. Let the spot get cherry red. The spot will rise and form a small dome. Quickly put the torch down. Hold a flat dolly firmly under the heated spot. Now take a flat-faced body hammer and tap the heated spot down. It is important to do this process very quickly, before the metal cools. When you strike the small dome, it should still be red. The idea here is to get the raised bump to compress or be compacted within itself. This type of shrinking takes some practice, but it works very well. You may need to repeat this process, moving around, to shrink a large high spot. Don't try to shrink the same small spot more than once or you'll run into trouble.

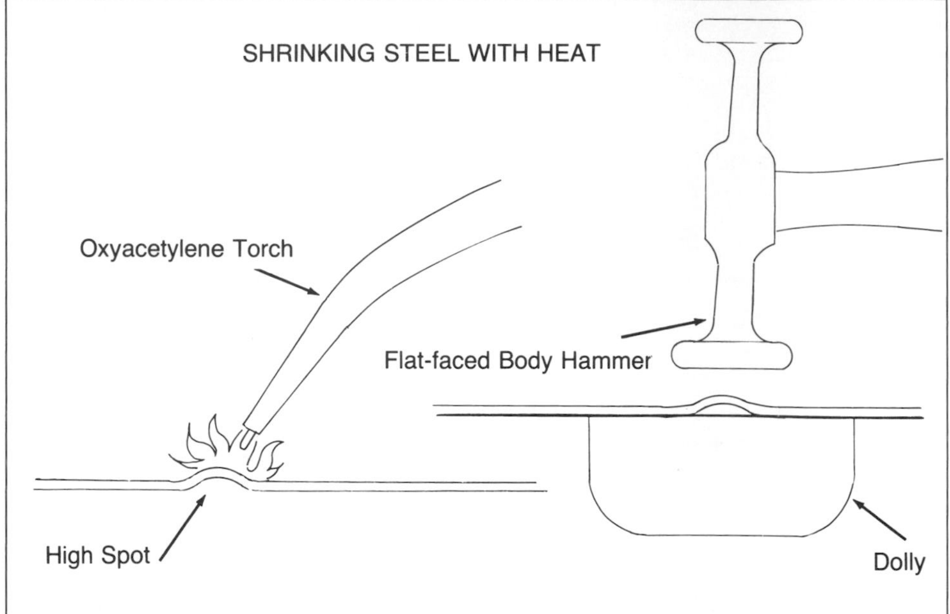

Shrinking steel with heat is done by heating up a high spot then hammering it down with a flat-faced hammer.

Marcel DeLey (left) and Lujie Lesovsky discuss the many uses of the Eckold Piccolo Kraftformer in auto restoration. These guys probably have forgotten more about metal shaping than most people will ever know.

After the area has been successfully shrunk with heat you must continue to metal-finish the area. Remember metal-finishing is a skill that takes time to learn. Don't rush yourself. Keep in mind your goal is to restore the surface back to its original *smooth* surface and shape without adding any fillers if you can help it.

Working Out Low Spots—Low spots are bumped up from the underside. This process is called "hunt and peck." You "hunt" for the low spot, and "peck" at it from underneath

Bill Bizer is using an Eckold Piccolo Kraftformer to repair a Duesenberg hood. The hood had a stretched area several inches in from the edge.

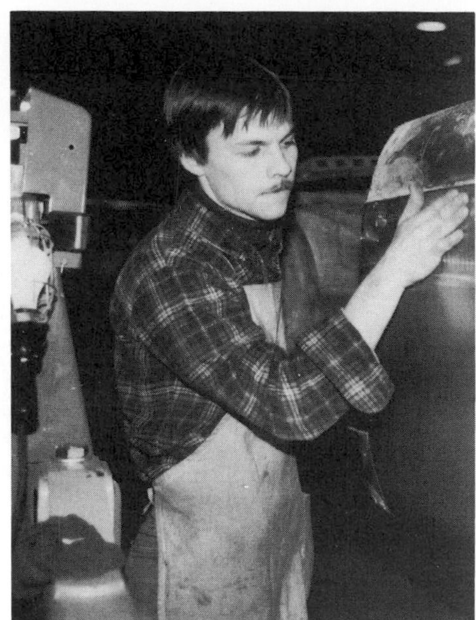

Replacing the whole lower rear fender of a Ferrari required careful fitting. A craftsman checked metal fit with both his hands and eyes.

with a hammer or other tool. As you bring up a low spot, run the file over it and more dye should come off.

There are many tools designed to raise low spots from underneath. There are several types of pick hammers and a special tool called a "bulls eye." These tools work wonders, though they tend to leave a mark on the underside of the metal. Depending on how critical the project is, these marks may or may not be acceptable.

Body men who make a living repairing wrecked cars see no problem with marks under metal. That's probably because they are only concerned with the look of the outer, painted side of the sheet metal. Hardly anyone looks under the metal on a common car driven on the street. Other body men, such as custom car builders and some restorers, wouldn't consider leaving visible marks on the underside of a metal part. They think of marks underneath as an amateurish imperfection.

My answer to this controversy is practical. I only mark the underside when it is absolutely necessary. I like nice, smooth undersides as much as any perfectionist, but I know sometimes it is very, very difficult to work without leaving marks underneath. I try to compromise and work as smoothly as possible. I have found you don't always have to bump up a low spot with a small, pointed pick hammer. There are other methods that work just as well.

Simply use a dinging hammer or a plastic mallet if the spot you are trying to move up is not real low. This works well for me because I use dye to metal-finish. The dye not only *finds* the low spot, it also shows the *size* of the low spot. I match the size mallet or hammer face to the size of the spot.

If the spot is wide, use a wide, soft-faced hammer with a soft blow. The wider the hammer used, the less mark it will leave under the metal surface. So choose the hammer carefully. The choice of hammer makes a difference in the end result.

Remember, it is important to restore the surface with a minumum of metal removal, because you must avoid thinning the metal. Keep the filing and grinding down to an absolute minimum, to keep as much metal thickness in the area as possible.

Scott Knight has recently invented a new tool for restoration work. It's called the *shrinking and leveling disk* (Shown on p. 101). It is used for removing small dents and metal-finishing with little or no loss of metal thickness.

FITTING BODY PANELS

Large body panels must be fitted and put into place when the body is bolted to the chassis. The body can easily twist when it is off the chassis. Don't do any major bodywork on a body while it is sitting on saw horses or jack stands. If you do, there may be big problems with the fit when it's time to bolt it back on the chassis. Doors that fit nicely when you started may not close, or may have gaps when they close, because the door jamb and hinge pillar are twisted. Quarter-panels may buckle as well.

This happens because the fitting and welding can make the body *firm* while it is in a twisted position. Let me explain. Suppose you weld two quarter-panels on the body while it is being supported by jack stands. The jack stands are just slightly uneven. The welding of the quarter-panels will strengthen the whole body, but will also reinforce the twisted, uneven position of the body on the jack stands. When you then try to put it on the chassis, the body will be twisted out of shape and won't fit. Such a twisted condition cannot be seen until it is time to reassemble the car.

Most old cars have many body mounts along the frame. The mounts

The lower 6-in. of this wheelhouse was rusted. We cut out all the rusted area and discarded it.

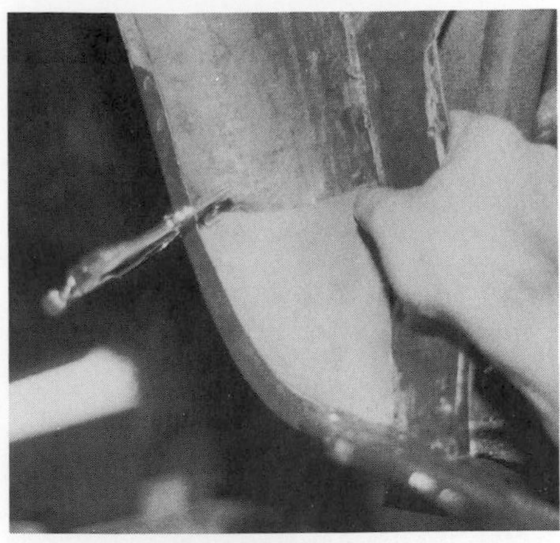

We made a patch pattern of chipboard. It worked well because it was flexible and could curve to fit the radius.

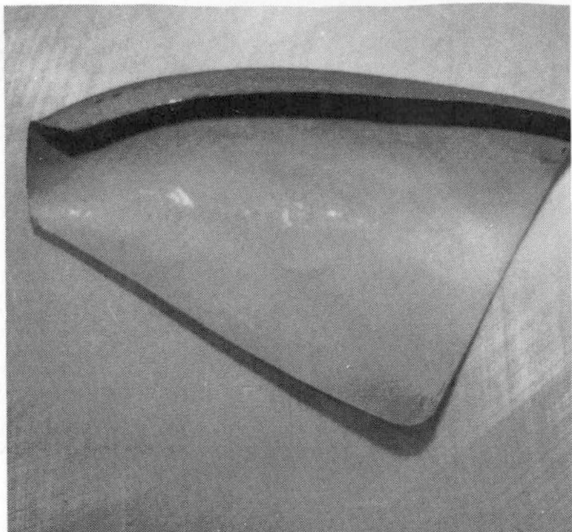

The patch panel fit like a glove. A tight fit is important for welding. A good weld ensures a good seal. We were able to repair the hole, but still retain much of the original sheet metal.

will most likely have shims or washers. When the major bodywork is done and you want to remove the body, be sure to mark the number and location of the shims and washers. This will really pay off on reassembly. Just mark them with tags describing their location, or number them sequentially around the car. It isn't a bad idea to make a quick sketch of the car and how you numbered the mounts and washers. Keep the sketch in a safe place, and refer to it when reassembling the car.

There was just enough of the old Ferrari floorpan left to tell us how to build a new one.

FLOORPANS AND WHEELHOUSES

Both *floorpans* and *wheelhouses* on older cars are often badly rusted. Some common makes of cars have replacement or reproduced parts that can be purchased and installed to replace rusted out floorpans and wheelhouses. Other makes of cars have no such replacement parts readily available, therefore these items have to be either repaired or made from scratch.

A *wheelhouse* is the metal part of the body that acts like an inner fender. It takes alot of abuse from rain, salt and mud, to name a few. Not all wheelhouses need to be replaced during a vehicle restoration, but almost all need to be repaired.

Be sure to do it right when you repair wheelhouses. Put back any beads the factory part had. Use the correct metal thickness, which is the same thickness as the original part. When you weld it in, use small welds. Grind all the welds smooth so they won't show on the final finish.

If it is necessary to replace the whole wheelhouse, be sure the weld goes *completely around* where it joins body and floorpan. The area where the wheelhouse meets the floorpan must be water tight. The seal between the seams has to be perfect.

Restoring floorpans can be tricky as well. They usually need repair due to rust damage. Try to make the floorpan look authentic. Don't let any welds show. Put in any beading or stiffening sections the floor may have had originally when it was built.

When mounting the repaired floorpan, use only original hardware. For example, solid steel rivets were commonly used to mount floorpans in cars of the Twenties, Thirties and Forties. Pop rivets weren't generally used in the automotive industry until the Fifties. If you're shooting for a completely accurate restoration, then you will need to use the original type of rivet used during manufacture.

The original floorpan may have been *spot-welded* in. If you have a spot welder able to reach where it is needed, fine. If not, you can simulate a spot weld by using a *plug weld*. A plug weld is made by putting a small hole in the panel—maybe 1/4-inch in diameter—then welding through the hole, filling it as you weld, to secure the panel to the mating body panel. This fills the hole and welds the part to the car. Just remember to use welding rod sparingly. Do not make the plug welds very large. The plug welds will

The whole trunk bottom of the Ferrari had to be replaced. All the stiffening sections and beads were duplicated in the replacement part.

look like *spot welds* if they are small, smooth and evenly spaced. If your plug welds leave a bump on top, smooth them off flush with the metal surface. A proficient welder should be able to make them smooth without any bump right off the bat.

Don't be afraid to use plenty of plug welds. The floorpan has to be strong and water tight. Using a *body seam filler* at this point is a good idea to ensure the seam is properly sealed.

Body seam filler is a special sealer, much like a thick paint, that is used on a body seam which isn't completely sealed by welding. After the floorpan is completely welded in and cooled, the body seam filler is brushed on the seam with an ordinary paint brush. It takes about two hours for the body seam filler to dry. Then it can be covered with regular paint.

BODY FILLERS

The two kinds of body fillers commonly used are *lead* and *plastic*. Although lead is more expensive and more difficult to apply, it is the more acceptable body filler in the restoration trade. Some people find plastic body fillers totally unacceptable. They think plastic body filler translates as "poor quality." This opinion developed over the years because a number of people have used the plastics incorrectly, gotten poor results and blamed the product rather than themselves.

My personal feeling about body filler is that I don't *want* to use either! I want to bring up all the low spots, lower all the high spots on the metal, and finish the job of restoration by priming and block sanding. Wouldn't it be great if that would always work! The truth is, no matter how hard you try or how skilled you are, there will probably be a low spot somewhere that can't be reached and worked out. For example, the end of a fender closest to a door usually is obstructed by an inner panel, or there is just no space to work at it from underneath. Therefore, it is necessary to use a filler. The cardinal rule to fillers is: *If you must use them, then use them as little as possible.*

Bill Bizer, a good friend and a highly experienced bodyman who has painted and performed bodywork on valuable antique and classic cars maintains that there is nothing wrong with body filler, as long as they are used correctly. Don't let someone convince you using a body filler is the sign of a poor restoration job. It is untrue. Bill's favorite response on this subject is: "Fillers are no problem if they are minimized and done *right*."

So the question becomes, how can you use body filler in the best way? There is a formula. You should spend alot of time metal-finishing first. Do *everything* possible to get the metal back to original condition and shape. Take the time to work carefully and completely. Spend as much time filling as you spent on the metal-finishing. Often the difference between a poor restoration and a great restoration is the difference between spending a little time and alot of time.

LEAD FILLER

Lead body filler has been used for nearly 60 years, both as part of the original fabrication of a car and for repair work. It may surprise you to know original bodywork on a production car included lead filler. For example, the place where the roof joins the quarter-panel of most cars is usually lead-filled.

It is not very difficult to use lead as a filler. It does take time and care to do it properly. Follow the steps and you'll get good results. Take shortcuts and you'll regret it, because every step in the leading process is equally important. Don't skip or skimp on any of the steps. Learning to make lead stick where you want it takes practice. Don't be discouraged if you have to try a couple of times to get it right. If possible, work on a horizontal surface at first. It is easier to do leading when the surface is level and horizontal. This way, gravity will be working for you, not against. Leading on a vertical, or angular surface is more difficult. Begin with a horizontal angle, then try more challenging surfaces.

Prepping The Surface—The first step is to thoroughly clean the area to be filled. Clean the surface to be leaded so it's free of rust, welding scale, paint and all other forms of contamination, or the lead filler won't take. The metal *must* be clean and well-prepared before applying lead.

A *grinder* can be used to clean steel. It cleans quickly and efficiently. A round wire brush on an electric drill also works well, especially to remove welding scale and to reach into places too small or too recessed for the grinder. Remember, if the surface is not

A rotary wire brush mounted in an air drill is just the tool to thoroughly clean steel before lead work.

Add lead filler to the tinning layer. Use it sparingly to avoid grinding work later.

cleaned properly the lead won't hold, and the area will lift or crack and eventually fall apart.

Tinning—After the area is thoroughly cleaned it must be *tinned*. Tinning is a process that prepares the metal surface to accept the lead. It is done with *tinning compound* or *tinning acid*. They prepare the surface with a very thin layer of lead to accept the lead filler later on. *Make sure that the tin-*

Heat the area to be tinned. Then spread tinning acid or compound on with steel wool.

Smooth the melted lead with a wooden paddle. A soldering torch tip—fitting over a regular torch tip—will produce a low, soft flame to make it easier to do leading.

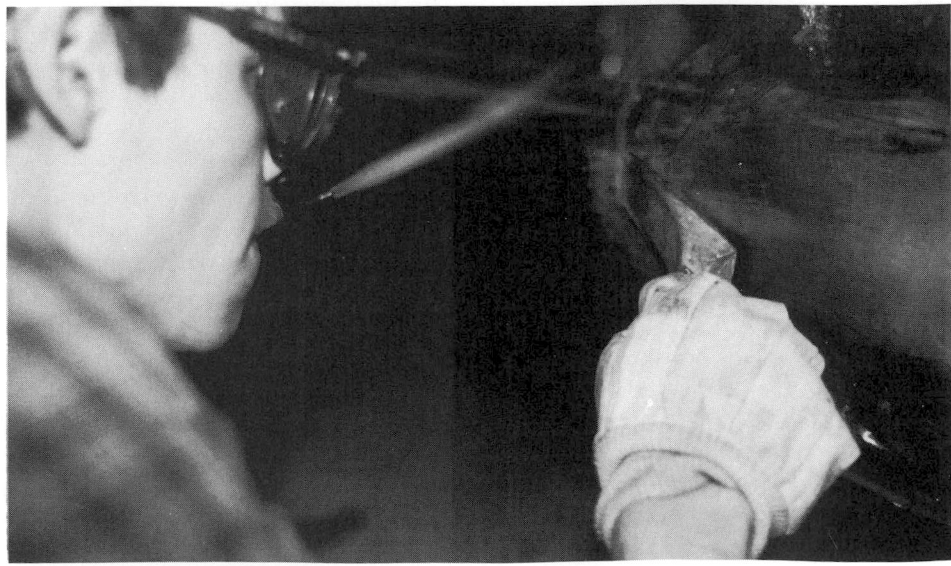

It takes a lot of skill and experience to use lead on a vertical surface. Too much heat will make the lead drip uselessly off the metal and onto the floor.

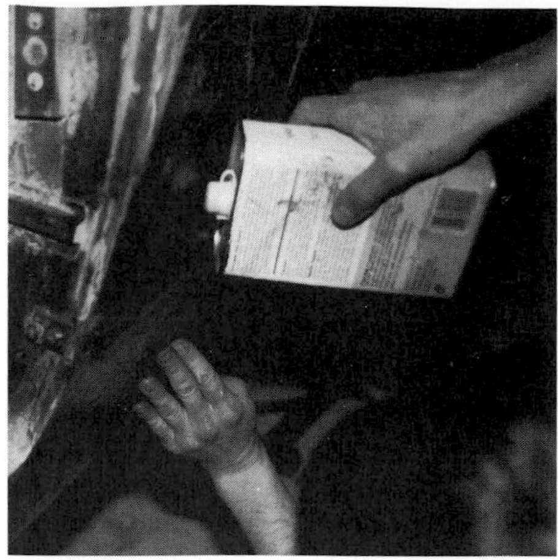

Neutralizing a surface after leading is extremely important. Any residue of tinning fluid or beeswax could make the final paint lift off the metal.

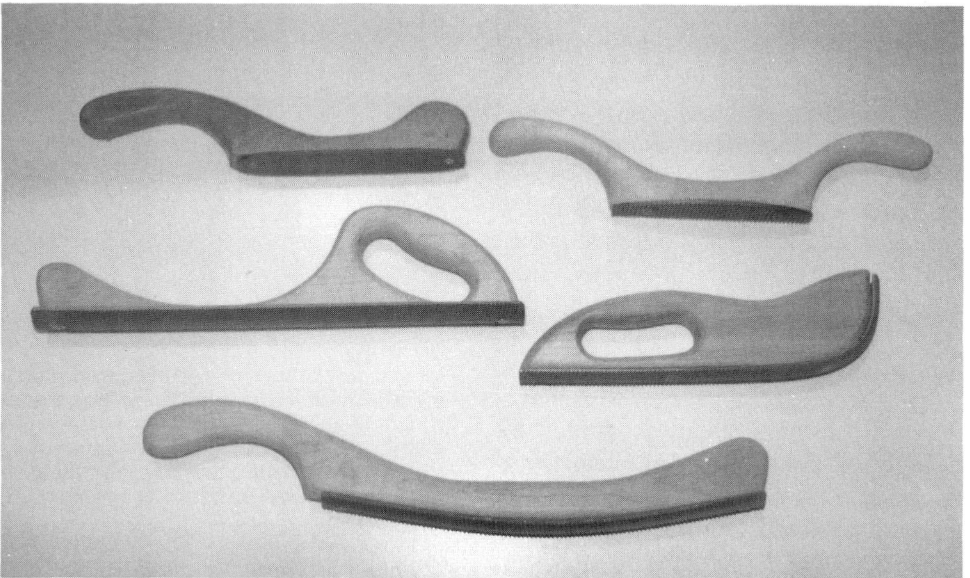

Here are a few of the body files available to use after leading or other body work. T. N. Cowan also sells a good rust remover to aid restoration work.

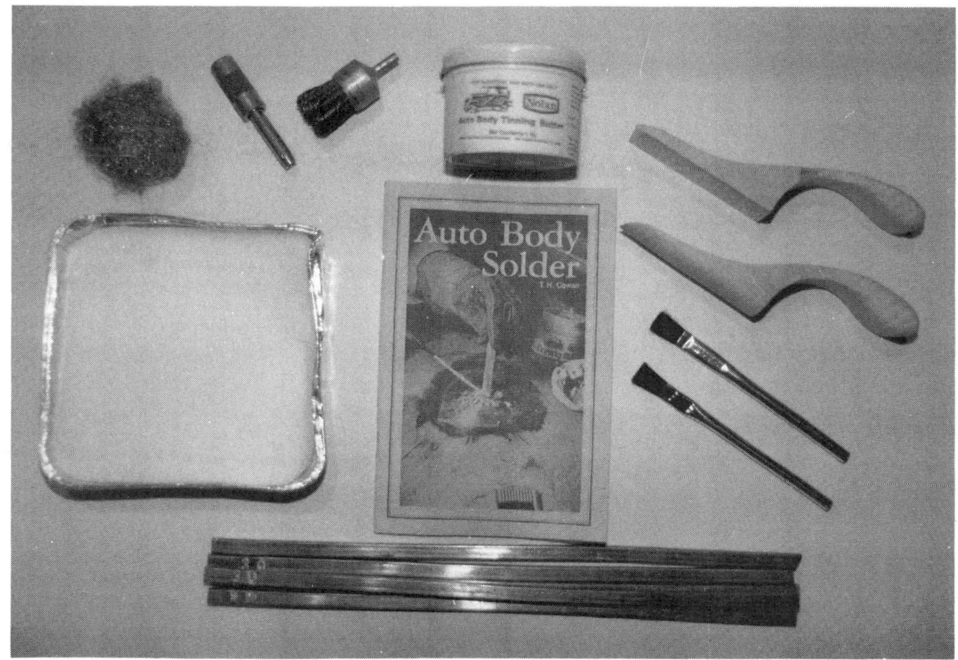

This book and these supplies for leading are available from T. N. Cowan. Terry Cowan's company offers good materials at reasonable cost.

ning compound or acid you purchase is for automotive use. Plumbers also use a tinning compound that isn't suitable for auto bodywork.

Use a welder's glove on one hand to protect yourself from the torch. Heat the surface area of the body panel slightly with the flame. Take care to avoid overheating the area or the surface will warp. Use a low flame on the torch and move it around gently to disperse the heat. Take a clean rag dampened with tinning acid or compound and wipe the heated spot with the rag completely, smoothing it all across the area you want to fill. No spot can be missed. If you don't spread the first, thin layer of base lead thoroughly over the area, the next layers could lift wherever a spot was missed.

This process is a bit difficult to explain. Lead, you see, will not adhere to other metal. It will only adhere, or bond, to itself. That's the reason behind the thin layer of tinning compound or acid. The heat from the torch opens up the pores of the base metal to accept the tinning compound, which acts as a *flux*. The steel body panel now contains lead that will bond with the lead filler to fill and finish the surface. This is the foundation for the actual filler layer of lead.

Lead Application—Lead comes in a bar. To apply the lead, hold the bar in one hand, and a torch set on a low flame in the other. Lead is only added to the area after it has been heated and tinning paste or compound has been applied. Only a *small portion* of lead is deposited onto the surface.

Once tinning is completed and the thin layer of lead has been fully applied to the area, you can start to *fill* with lead. Start by heating the lead bar and the work area. Then melt the lead directly onto the metal surface until you have an adequate deposit of lead, about the size of a half dollar and 1/4-inch high. Keep spreading heat on the work area. Use just enough heat to keep the lead melted, not so much that it liquefies. You'll soon get a feel for how much heat that takes—too much and you lose control of the lead and it starts dripping on the floor.

Working The Lead—Spread the lead around with a wooden paddle, which can be purchased from a body shop supply store. Use *beeswax* on the paddle to keep the lead from sticking to the paddle itself. Beeswax is also available from body shop supply stores. If the lead has hardened a bit, then you may apply a very low flame carefully over the area just until the lead softens. If you heat too much at this point and an area falls out, then reapply more lead. When you have deposited as much lead as needed to fill the low spot, spread the lead with the paddle, as smoothly and as evenly as possible. Lead solidifies very quickly, and you will be able to begin filing almost immediately.

Filing—Go over the surface with a body file, until the high spots are removed and the area feels level. Continue checking with your hand by rubbing it across the surface periodically. Some people have a tendency to overfile lead, and if you do this, you'll have to start over. After filing, use 80-grit sandpaper wrapped around a wooden block and sand the lead-filled area until all of the file marks, if any, are gone. If you file carefully, you shouldn't have any marks at all.

Cleaning—After the lead has been applied, filed and metal-finished, the area must be carefully cleaned. Leading compound and/or acid should be completely neutralized. This is done by washing it with hot water. Adding a small amount of either vinegar, ammonia or sodium bicarbonate to the water will help neutralize the flux. To be sure all the beeswax is removed, use *metal prep*. Metal prep is a special cleaning fluid used to prepare and clean the surface of metal to be painted. You can buy it at body shop supply places.

Remember auto body leading is a skill that requires a great deal of time and patience to learn. You aren't going to pick it up in five minutes. I recommend you find an old fender or some other discarded body part and use it to practice on. Leading is a skill you'll need often if you do alot of sheet metal restoration. All the supplies you need should be available at any large auto body paint store.

Some tool suppliers, like Snap-On and Mac, also carry a limited supply of leading tools. The main items they carry include beeswax, vixen files and wooden paddles.

When purchasing the body lead or *solder* (which is its common name) be sure to ask for *30/70* lead. This means the bars are made of 30-percent tin and 70-percent lead. This combination works best on auto bodywork and is used by most bodymen and restorers.

I recently found a comprehensive

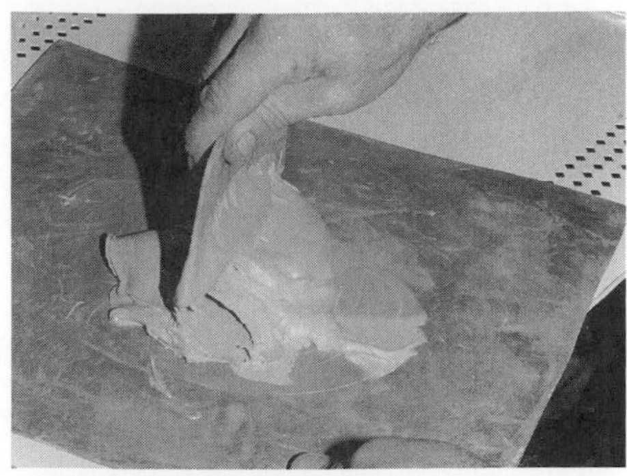

Use mixture of catalyst and filler according to manufacturer's directions. Generally, a good rule of thumb is 1/2-inch strip of catalyst for every golf ball size of filler. Mix thoroughly, careful not to overlap and trap air bubbles in filler. Air bubbles may lead to failure of the filler later on.

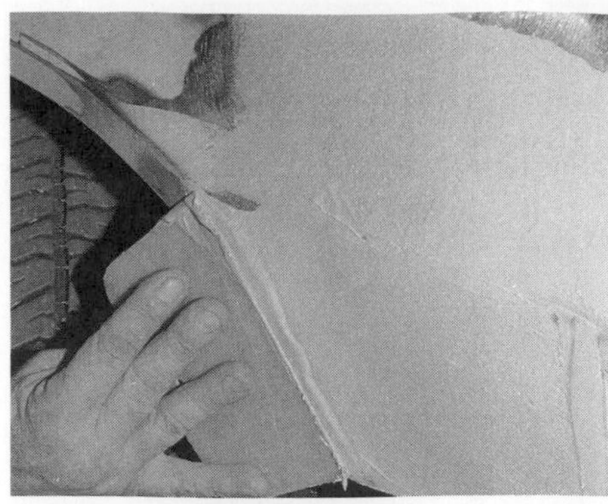

After grinding and cleaning area to be filled, George of GKR Custom Paint & Body in Chatsworth, California, applies filler with a plastic squeegee, moving in one lateral direction only. Avoid overlapping layers which can create air bubbles.

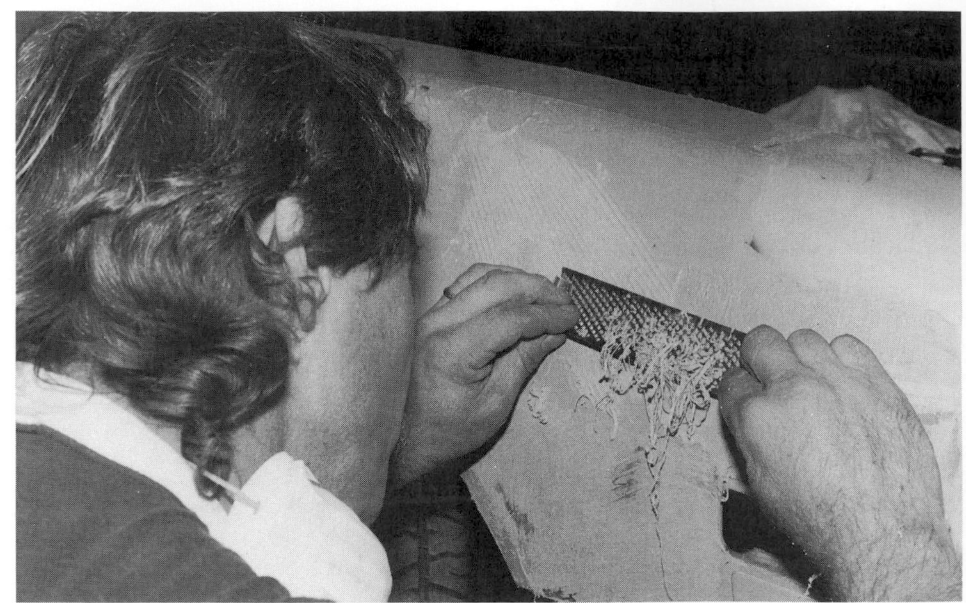

Once filler "kicks over," George uses a Surform file to cut down the excess. If the filler crumbles, it's too hard, and should dry more.

LEAD FILLER SAFETY TIPS

1. Always do lead work in a very well-ventilated area because the heated leading compounds and acid give off harmful toxic fumes. Even if you get chilly from opening windows or doors while leading, it is better to feel cool than to inhale poisonous fumes.
2. Always use files to remove lead. Never use a power grinder or sander. Grinders make lead dust that is poisonous and hazardous to your health if you should swallow or inhale it. Files make lead shavings, which fall harmlessly away.

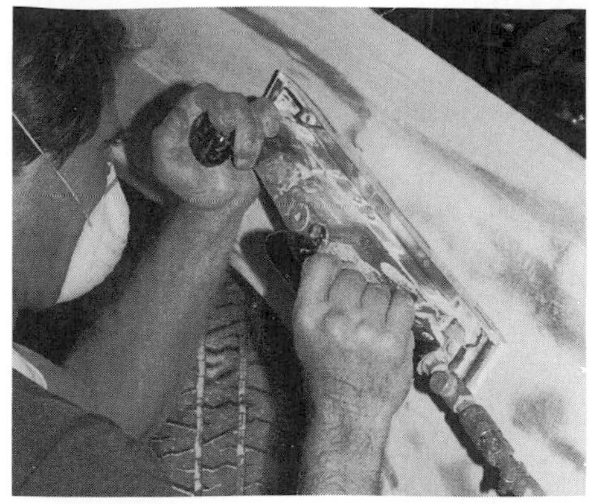
An air file equipped with 150-grit sand paper is the easiest way to go for the next step. If you don't have the fancy tools, then, a sanding board will have to suffice. Note here that George is sanding roughly 45-degrees off the horizontal. Sand in this direction, then flip file around and sand in other direction at same angle. This is called "cross-hatching."

Checking for contour is best done by rubbing hand across panel. It takes awhile to develop this "feel."

Final step is to use a D/A (dual-action) sander with 220-grit sand paper to prep the surface for primer.

book on auto body lead work. It is *Auto Body Solder* by T. N. Cowan. I highly recommend this book to anyone who wants to read more about the specialized process of leading. You can order the book by mail by writing to: T. N. Cowan, P. 0. Box 900, Alvarado, TX 76009. The book is clear, well-written and fully illustrated. It also tells where to get everything you need to do lead work. It includes many safety tips. If you are attempting lead work for the first time, this book is a must.

PLASTIC FILLERS

I honestly feel there is nothing wrong with plastic fillers, as long as they're used as they were intended. Too often, amateurs use plastic filler in place of hammering and dollying. Sometimes it is used to patch up rust holes, which it was *never* intended to do. When plastic filler is used in this way, it will absorb moisture and become loose. That's when problems start to occur.

Another reason I hesitate to use plastic filler on cars built before 1955 is one of authenticity. Pre-'55 cars just didn't use it because it wasn't available. However, if you do decide to use plastic filler, buy the best you can. Don't bargain shop. The best product will prove to be the cheapest in the long run. Then take care to apply it the right way.

Like lead, plastic filler needs to be applied over solid metal which has been thoroughly cleaned. The metal cannot have any holes in it. Be sure to make the surface as even and smooth as possible before you start using filler, such as with a grinder.

Applying Filler—You should only mix enough plastic filler to give you a 1/8-inch thick layer. As a general rule, for every "golf ball-size" portion of filler, mix a 1/2-inch long ribbon of catalyst. Mix with straight strokes rather than overlapping ones to avoid air bubbles. This ratio should only be used as a guideline. Weather, temperature, and age of materials will determine the exact ratio. The correct hardening time allows you three or four minutes to apply the material. If it hardens too fast, adjust the ratio of filler/catalyst to get more time.

With a plastic or spring-steel spreader, spread the filler over the repair area. Use long strokes *in one direction only*. Don't change directions as you add and overlap filler, or it will lift the filler underneath. Fill the area using as little filler as possible.

Filing Filler—If the catalyst/filler ratio is correct, the material should be hard enough to file in 5 to 10 minutes. Most professional bodymen prefer an 8- to 10-inch curved Surform file blade because it offers better control and works on both concave or crowned areas equally.

When filing, draw the file over the surface, cutting off thin curls. If the curls crumble or clog the file-blade, the filler hasn't cured enough. Wait a few more minutes, but not too long. Rather than pushing the file, draw or pull it for better control. As with lead, use your hand to check for high spots, and don't overdo it by filing down to bare metal. When this draw filing is done, use a 36-grit paper and block sand the area using a *cross-hatch* pattern. Cross-hatching is done by drawing the sanding board across the filler at a 45-degree angle to the imaginary vertical line of the area, then doing the same on the opposite side.

Concluding Thought—It is beyond the scope of this book to cover *everything* involved in restoration. That information would require several books, and it just so happens that HPBooks also publishes a few that go into greater detail. They are: *Metal Fabricator's Handbook; Paint & Body Handbook; Welder's Handbook*; and if you're restoring a pre-'49 model, you might find HP's *Street Rodder's Handbook* useful as well.

CHAPTER

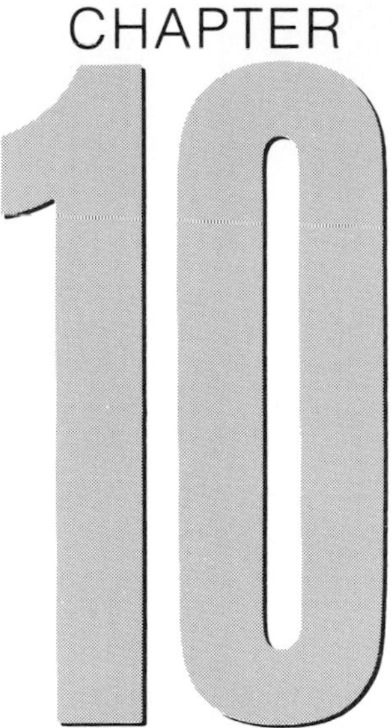

SHEET METAL INTERIORS

Many custom cars and nearly all race cars have specially-built sheet metal interiors. An interior can be very complex: a complete car interior that includes the floor, wheelhouses and firewalls. Or it may be something simple that involves nothing more than a couple of door panels. The extent of the complexity of the interior is dependent on many factors, such as budget and type of car, racing regulations, and what the car will be used for. There are, however, basic fundamentals and a procedure that apply to all sheet metal interiors.

DRAG CARS

Many drag cars, such as those in Pro Stock and Gas classes, use *full* aluminum interiors: I use 0.050-in., 3003-H14 sheet. It is easy to bend and shape, yet durable. Also, 3003-H14 is lightweight, easy to weld and it anodizes well, taking well to color or clear finishes.

These interiors include every metal or trim panel normally found inside a passenger compartment or engine compartment. From front-to-back, the interior is totally hand-fabricated—a custom package. Custom aluminum dashboards, or instrument panels, are even found in altered classes.

STREET MACHINES

Street machines (the current term that has all but replaced the words "hot rod") have been on the rise. Though the idea of modifying a classic car is still the same as it was back in the Fifties, the technology and choice of car have changed. Back in the Fifties, the Thirties Ford was the car of choice.

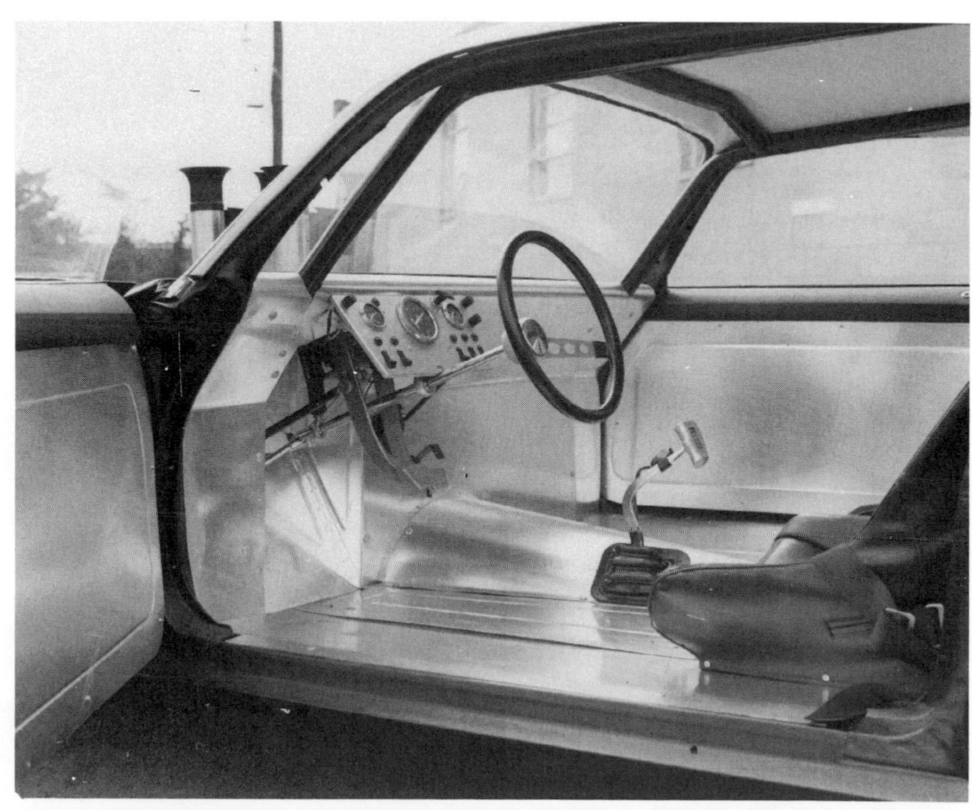

A car must be fully constructed before you can build the interior. After the interior is in place, it is extremely difficult to make a roll bar change or relocate the steering column.

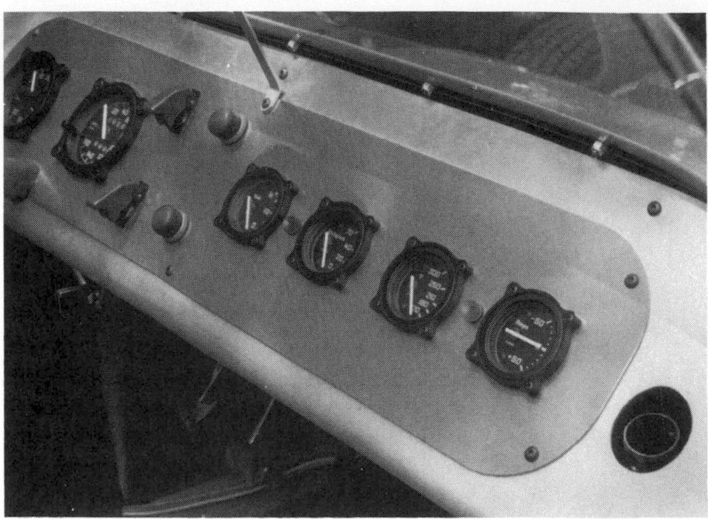

Here's the dash panel I constructed for my street rod. You must measure all gauges and switches before building the panel.

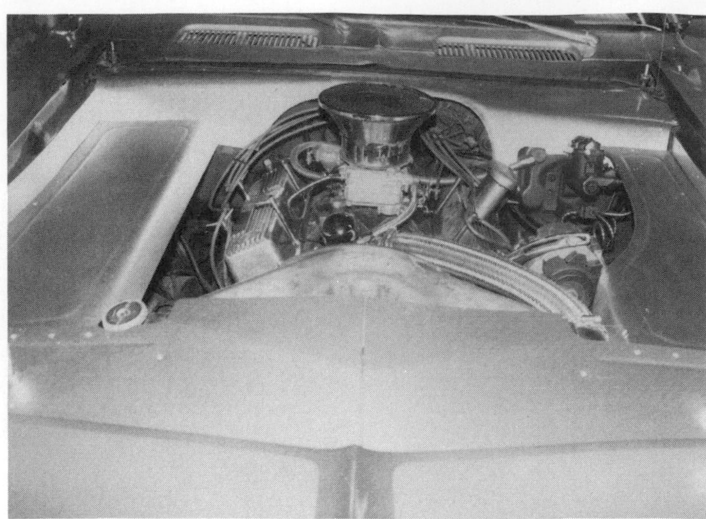

Engine bays lend themselves nicely to sheet metal work.

Far more popular today however, are the cars of the mid-Fifties and Sixties.

The concept of a street machine is only limited by the imagination. Some cars feature so much custom work that they are labeled as a "pro street" car, which is essentially a "street-legal" version of a Pro Stock drag racer. Most all pro street cars feature an extensive, hand-fabricated aluminum interior, with huge aluminum wheel tubs to accommodate 18x33-inch rear tires, transmission tunnels, floorpans, engine compartments and customized instrument panels.

Most sheet metal on today's street machines is custom-made. The trunk of this Camaro features wheelhouses, rear firewall and fuel cell—all fabricated by hand. Photo by Jim Kelso.

ROAD RACING CARS

Road racing cars, such as the Grand Touring cars or sedans, also use aluminum interiors. These interiors are generally not as extensive as the ones used in drag cars. IMSA and SCCA rule books spell out which stock panels must be retained. Stock floorpans are mandatory in some classes. Other road racing classes have more flexible or lenient rules, such as IMSA's Grand Touring Prototype (GTP), where it is legal to use a *complete* aluminum interior to replace the stock one.

CIRCLE TRACK

Stock car interiors are usually a combination of steel and aluminum. Steel components are used for critical areas such as floorpans, and front and rear firewalls. Because it resists heat better, and is generally stronger and

Today's race cars feature completely custom-made interiors. Rules differ greatly between race organizations. Be sure to check yours out thoroughly. Photo by Michael Lutfy

This GT racer retains stock floorpan as required by rules. Custom-fabricated aluminum instrument panel is anodized blue.

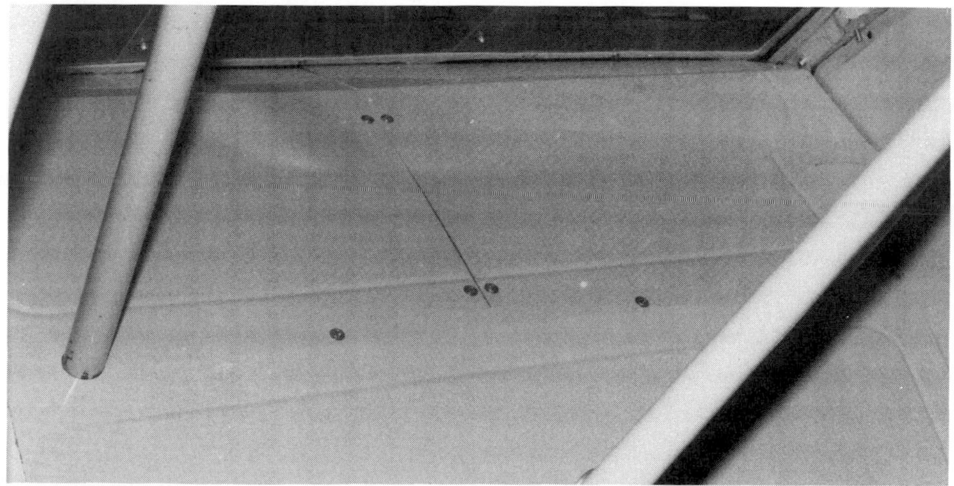

Interior panels in and out of GT car. Display of interior panels shows simple, but effective beading and mirror-image right and left parts. Panels fit tightly to each other and around roll bars. Dzus fasteners hold them tightly in place.

more crash-resistant, it is best to use steel in those areas: 20-gage (0.0359 in.) minimum. Aluminum can be used for the rest of the interior panels to save weight, which is important for stock car racing.

Rules—There are many different circle track sanctioning bodies, each with its own set of requirements. Each may specify what types and thicknesses of metals can be used for specific interior parts. It is vitally important to obtain the rules book from your sanctioning body. Study it. Follow the rules. If you are in doubt, be conservative. *Always use the material that will make the safest interior possible.*

The rules will determine to a large degree how extensive the sheet metal interior work will be. You will also get an idea *why* the interior is regulated. It may be for safety reasons, or to ensure that you don't reduce weight illegally. Frequently, certain stock sheet metal panels must be retained to give the "illusion" of a stock-based car. The idea is to have safe and competitive cars racing against one another, and to retain some "product identification" for marketing reasons.

BASICS

Whatever type of car you're working on, some general guidelines will apply to its interior. A sheet metal interior must protect the driver and should be attractive. An interior doesn't have to be ugly to be safe. Nor does it have to sacrifice protection to be pretty. It can serve both functions if the design is well-planned and all panels are built and installed with care. Let's look at some basics on how to accomplish *both* goals.

Driver Protection—The first goal of a sheet metal interior is to protect the driver. What *kind* of protection can an interior give a driver? The interior shields the driver from three potentially dangerous elements: fire, fumes, and flying debris.

A protective metal interior must be of the right materials, fit well and be durable. Rule books usually specify the metals for certain areas, and how thick they must be. For an interior to be durable, it must be carefully designed,

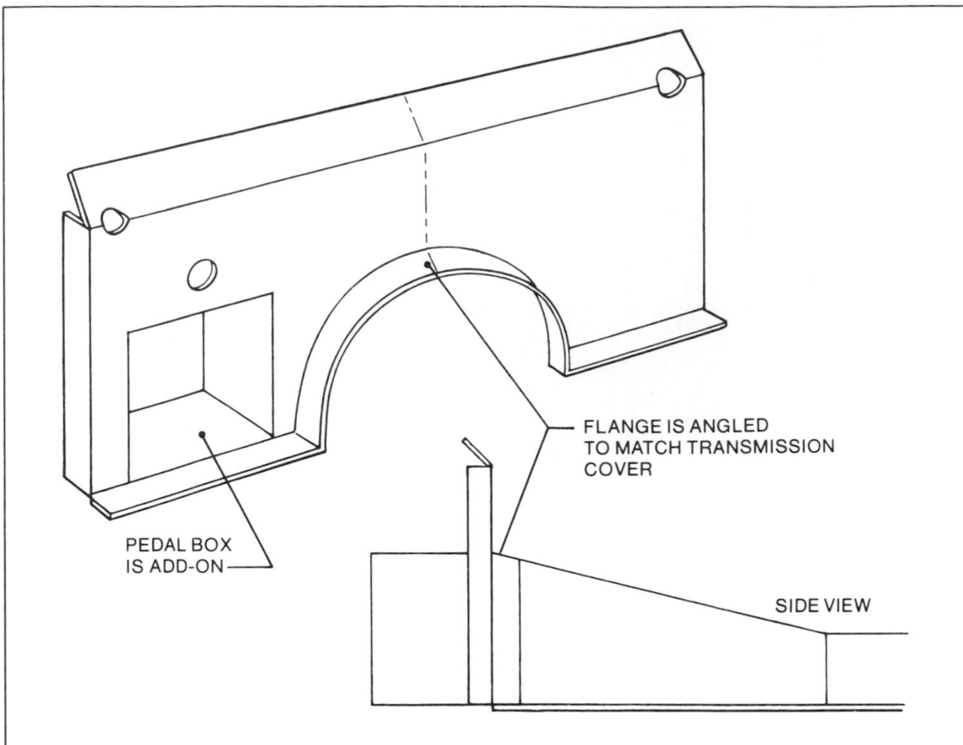

Typical firewall has some or all of these items: holes for roll bars, flange for transmission cover, floor-mounting flanges, steering column hole and pedal box. Make a sketch and take measurements when starting development of a firewall.

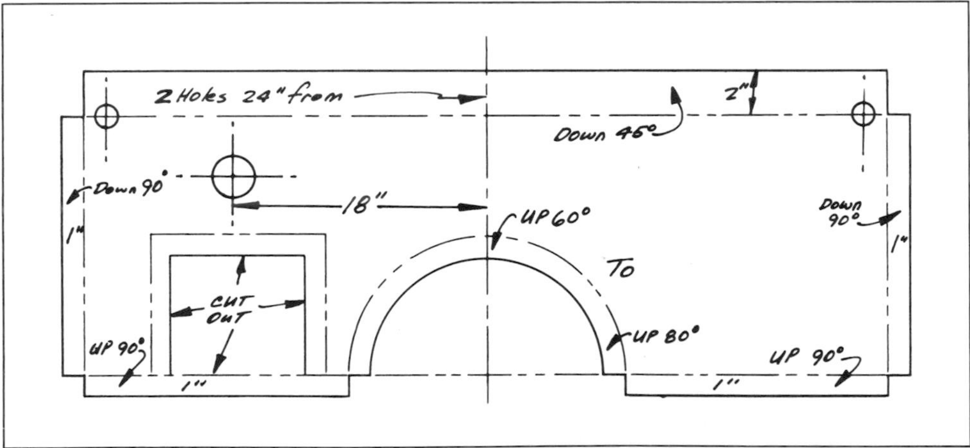

Transfer all measurements and information to a full-scale layout. Be sure to include all information: flange to be added, bend lines and amount or direction of bends.

fabricated and installed. Panels must be fitted tightly to each other and the car. Take great care to prevent gaps and misfits. Similar to a leak in a boat, a gap or misfit will keep an interior from sealing the way it should. Remember: *One small gap can let in enough fumes or fire to endanger the driver's life.*

The *installation* of the interior panels can be the most critical step in the whole process. Not only must they fit well, they must be *tightly* attached with the right type and number of fasteners. A loose panel is just as bad, if not worse, than no panel at all. Usually a loose panel will not come off completely. However, if a corner pulls loose in an accident, it may become a giant knife blade and cut the driver. It can also become an open door for burning oil or fuel. No matter how you look at it, a poorly installed interior can be extremely hazardous.

FRONT FIREWALL

A front firewall is usually the most difficult interior panel because of the many components it must accommodate: engine and transmission (with a front-engine car), steering, driver and driver controls, frame and roll cage, exterior structure and panels.

Making The Pattern—The procedure used to develop a pattern for a front firewall is identical to that used to develop *any* interior panel pattern. Follow the same procedures for pattern development, metal cutting and forming, fitting and installation when making the remaining interior panels.

The engine, bellhousing and transmission must be in the car before

INTERIOR HINTS

1. Fit each panel to the car and to the next panel as you go. This allows you to get a gap-free interior more easily.

2. *The order for making and fitting interior panels is very important.* Following a certain order works best. Because most panels must fit to other panels, it is necessary to make some first, others second, and still others later. Save yourself some grief. Build according to the order I describe.

3. Start with the front firewall. It is often the most complicated panel. Take time to develop this panel and make sure it is right. In many cases the front firewall includes a wide lower flange where the front edge of the floor rests. It also includes a flange around the transmission cover.

4. When fabricating an interior, don't overlook the fact that you're building a *complete interior,* not a bunch of unrelated panels. So keep the "big picture" clearly in mind as you go. An interior must have *continuity:* All panels must combine to be functional, simple and attractive. Use large-radius bends wherever possible to soften the visual impact and create a beautiful interior. Avoid sharp edges and bends when you can.

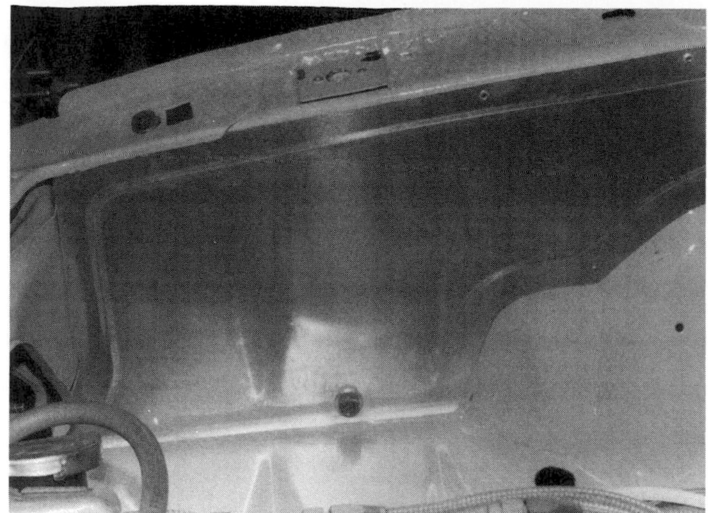

Bead simply follows the periphery of the panel. This type of bead is mainly for appearance.

Tube construction allowed me to make most interior panels before the body was installed. Inboard wheelhouse panels are hammer-formed. Photo courtesy Steve Allen.

you can fit the front firewall, if it is a front-engine car. The same applies to a rear-engine car when doing the rear firewall. Start with the basic dimensions. Measure across the front firewall area from door pillar-to-door pillar. Measure up from the floor, or where you think it will be, to the underside of the windshield cowl. Construct a rectangular-shaped pattern from a big piece of cardboard using these two dimensions. If you don't have a piece that big, tape smaller pieces together. Draw a vertical center line on the pattern.

Locate the bellhousing opening. Measure from the top of the cowl to the top of the bellhousing. Transfer this point to the pattern. Determine the bellhousing radius. Find its center and draw it on the pattern. Add 2-inches to the bellhousing radius to provide bellhousing clearance. Snip out the bellhousing opening from the pattern. This opening will have to be modified later to allow for the transmission cover flange. Fit the pattern over the bellhousing and check for accuracy. If there are any gaps or fit problems, correct the pattern.

Pedal Box—Depending on engine or driver location, your firewall may need a *pedal box*—a protrusion of the firewall that extends into the engine compartment. The pedal box accommodates the clutch, brake, accelerator pedals and the driver's feet. It will be a five-sided box attached to the engine side of the firewall. The pedal box must be large enough to clear the pedals and allow for comfortable foot and leg movement. The bottom surface of the pedal box should blend in with the floorpan.

After you've developed a pedal box, mark its location on the firewall pattern. Add 1-inch flanges to the firewall panel for attaching the pedal box. Indicate the direction that you'll want the flanges bent. *Make the marks on the side the flange will be bent toward.* Bend the pedal-box flanges toward the engine compartment.

Pattern Details—The front firewall pattern must include openings for components that must pass through it, such as the steering column or roll bar braces. Each opening must match the size of whatever part fits through it and it must be located *exactly* where the object will be.

Cut a slit from the edge of the pattern so you can fit it around parts you're cutting holes for. While you are pulling the pattern in-and-out of the car to make and cut holes for the steering column or roll bars, fit its periphery. The pattern should be marked and trimmed to match the areas you'll mount it to.

Make the firewall pattern a map or set of directions. For instance, to add a 1-inch flange at the bellhousing opening, write ADD 1 IN. on the pattern. This flange will be bent toward the passenger compartment, so mark on the passenger compartment side. A 1-inch flange is needed at the bottom of the firewall to support the leading edge of the floorpan. Indicate this on your pattern too.

When transferring the pattern to metal, add the flanges and transfer the instructions as well. When the metal is cut, you'll have the extra 1-inch you need for the flanges.

This system will develop a simple, clear, cardboard pattern and a set of instructions so you can construct a complicated metal panel. All you have to do is remember to make your notes *while* you're developing the pattern. Afterward, be careful to read them before you cut or bend the metal.

Don't be surprised if you end up scrapping the pattern and have to start over. But first, follow standard practice by patching the openings with taped-on cardboard pieces until the pattern fits. You won't have any problems telling when it fits. The pattern will slide into place and fit tightly to all the mating surfaces and components without bulging. When the pattern fits well, remove it and take it to the work bench to transfer the information to the metal blank.

CARE OF THE METAL

If you want a scratch-free aluminum interior, use aluminum with a stick-on

ANODIZING

Although anodizing can be done in a range of colors, colors tend to fade from exposure to sunlight. I recommend a clear anodized finish for all interior metal panels. It protects the aluminum from oxidation, looks good and won't fade. Without clear anodizing, aluminum panels oxidize to a dull, dreary gray color. They lose their original shine. Considering the investment you'll be making in those interior panels, it only makes sense to have them anodized. Never send an aluminum panel to an anodizer with a steel rivet or fastener in it. Steel makes anodizing impossible.

cover. Although it costs more than bare aluminum, it is worth it. There is less chance of damaging the metal surface as you work with it. As a bonus, the plastic or paper stick-on cover is easy to mark. Keep this cover on as long as possible. Much of the metal work, such as bending, beading and drilling, can be done without removing the cover. To weld, merely peel back the immediate area of the cover approximately 6-inches from the weld line. When you've finished welding and the metal has cooled, lay the protective cover back in place. It may not fit perfectly, but it will protect the surface as you continue to work.

Anodizing is another reason to use sheet aluminum with a protective cover. If you intend to have the aluminum panels anodized it is important to keep them unmarred (see sidebar).

FROM BLANK TO PANEL

Marking the Blank—Lay the firewall pattern on the sheet metal. Mark the metal or its protective cover with a soft black pencil. Follow any instructions marked on the pattern. *If you want to bead the panels, this must be indicated on the pattern.* You should decide if and where you want to bead the firewall. Remember: A beader's throat-depth limits how far a bead can be put from the edge of a panel. Mark the beading instructions on the metal blank.

Beading—Beading serves two purposes: It stiffens a panel so it flexes less easily. It also enhances the looks of a panel. Just remember not to get carried away. A little beading can go a long way. It should be simple and consistent on all interior panels. I've seen a lot of overdone beading jobs. They made me think a seven-year-old kid got hold of a beader and thought it was a neat toy.

Cut out the metal blank according to your instructions on the pattern. Once it's cut out, the blank is ready for beading or *stepping*. Stepping is a means of forming overlapping joints between panels without unsightly raw edges showing. It also keeps the surfaces of the two panels flush with one another.

Stepping—Stepping is done with a special set of beader dies. You either make them yourself or have them made according to how deep a step you need. The step should be as deep as the thickness of the overlapping panel. So if the overlapping panel is 0.050-inches thick, the step must be the same: 0.050-inches. Stepping, like beading, must be done before any other forming operations. This order has to be maintained for a simple reason. Folds or bends will interfere with the beader shafts. You'll have unbeaded areas near any fold if you don't bead or step the blank *before* bending.

Flanging—Bend flanges after beading or stepping. Straight flanges can be made in a sheet metal brake or over a straight edge. Curved flanges, such as the one at the bellhousing opening, can be made by hand. Bend the flange with glass pliers, hand seamers or a hammer and dolly. Any of these will do.

Fitting the Panel—Final fit each interior panel until it fits perfectly before going on to the next one. Each edge must match up to its mounting. If the panel doesn't fit, snip or file it until it does. If the panel is too small you can do one of two things. Scrap it, or add metal to it. Neither is desirable. Be sure you allow a little leeway in cutting the panel in the first place. Don't be like the guy who said, "I don't understand. I cut it twice and it's still too short." Sounds funny now, but not when it happens.

Check all clearances before building a wheelhouse tub. Tire is held in its full-up position. Added clearance is needed at the end of the rocker panel.

Installation—When you have it right, Cleco the firewall in place until it needs to be removed to fit neighboring panels. Mate each panel to the adjacent panels. Only when all the panels are formed, fitted and anodized should you final-install anything. Once all the panels are in place, they should match perfectly. It is not uncommon to find an interior with dozens of pieces. Matching them right is no small accomplishment. Take the time to do it right!

WHEELHOUSES

Wheelhouses, or *tubs* are next. They must be in place before a rear firewall or floorpan can be made. Here are some important factors to consider while developing wheel tub patterns.

Measurements—Clearance must be provided for the tires as they move up and in with the rear suspension. Drag cars need additional rear tire clearance because they grow—their outside diameter increases as the tires spin. Unless the rear wheel tubs allow for this growth, the tires will rub against the wheel tub and may cause tire failure, which may result in a serious crash.

If your car is a drag race car, find the *expansion factor,* or how much the rear tires will expand. The tire manufacturer can supply this information. Use the expansion factor in combination with the other tire and suspension

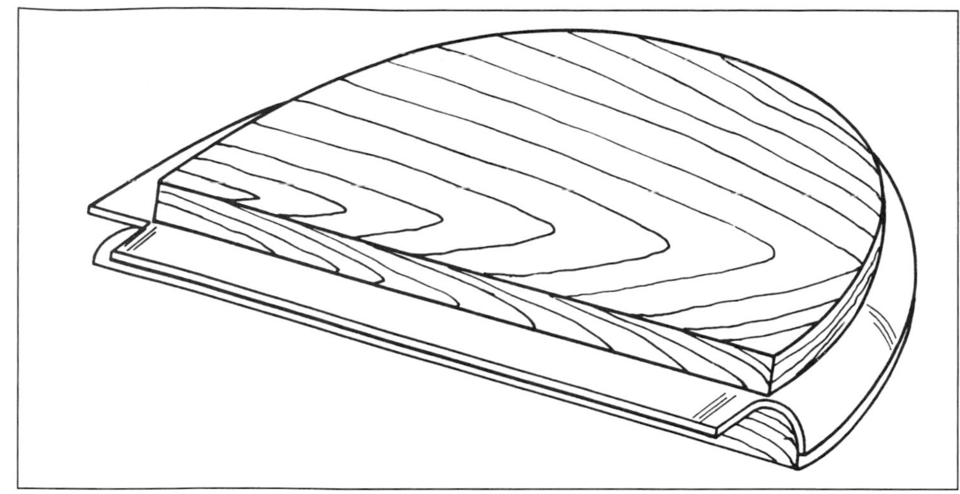

Simple hammerform is worth making. Use one to produce a beautiful, professional wheel tub.

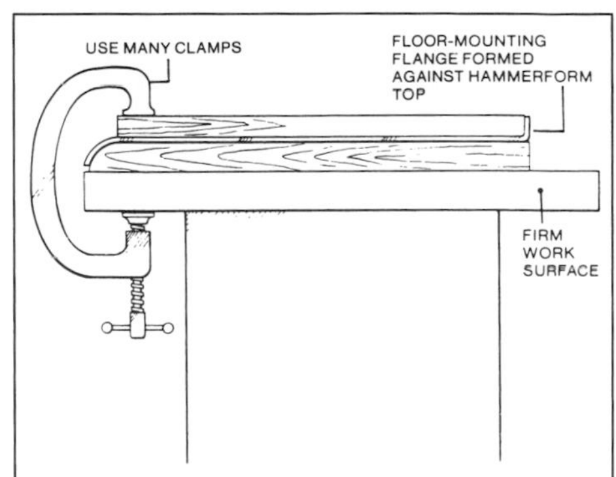

Although only one C-clamp is shown, use several. Clamp blank securely to prevent it from shifting during hammerforming.

measurements to construct rear wheel tubs that will provide sufficient tire clearance to avoid tire rub.

For example, suppose the tire diameter is 30-inches "tall." If its diameter expands 3-inches, you must allow 33 inches for the tire. Add further a 2-inch clearance in front and in back of the tire. The width of the wheel tub—front-to-back—should measure 33 in. + 2 in. + 2 in. = 37 in. Allow for suspension travel at the top of the wheel tub. No two cars have the same suspension geometry or wheel tub. To determine with accuracy what clearance you must allow at the top and sides of a rear tire, run the rear suspension through its travel with the wheels and tires installed. Remove the springs (unless it's a standard leaf-spring rear suspension) to make the job easier. Measurements must include the width of the tire plus clearance on the *inboard* side of the tire. Make a vertical section template and check it with the tire all the way up. Position the template to a "hard" point such as the frame rail. Match-mark the template with "witness lines," so it can be accurately repositioned for checking the wheel tub.

In some Gas and Pro Stock drag cars the width of the tires can be a problem. The frame itself must be designed and built very narrowly. Otherwise you'll never be able to enclose the tires under the body. If the frame was not specially built or designed for drag racing, it may have to be narrowed before fabricating the rear-wheel tubs.

Wheel tubs aren't hard to build. Just follow some basic guidelines. First of all, keep the wheel dimensions and clearances in mind and write them down. Make templates from these figures. Trim the cardboard pattern to fit against the inside of the quarter-panel and frame, then tape it in place.

Shape—The shape of a wheel tub is simple. A semi-circular side panel joined by a curved top: like a 55-gallon oil drum cut lengthwise in equal halves, the halves cut in equal lengths. Here's a simple method to make a wheel tub:

Make a semi-circular pattern for the side. Use this pattern to make a sheet-metal blank. Cut and roll a long rectangular strip of metal wide enough to enclose the tire, plus some clearance. This piece curves around the tire. Weld the curved piece to the side piece and the result is a functional wheel tub. This is a simple method, but I use a different one.

Hammerforming—makes for a more professional-looking wheel tub, and it is the method I use. You still need a full-size pattern for the side and for the long strip. Make the pattern for the side according to your measurements. Measure the periphery of the arc of the curved side. The rectangular strip is as long as this arc, plus additional length for front and rear mounting flanges. The width of the strip is determined by tire clearance on the inside, plus tire width and the distance from the outside surface of the tire to the quarter-panel.

Here comes the trick. Transfer the pattern of the side of the wheel tub to a large piece of 3/4-inch-thick plywood. Cut it out. This plywood piece is your *hammerform base*. Grind a *smooth* 1/2-inch radius around *one* of the curved edges. The opposite edge should remain just as you cut it—square. Leave the straight edge at the bottom alone.

Make the upper half of the hammerform from another piece of 3/4-inch plywood. Trim it about 1-inch smaller along the curved edge. Use this as a top piece to secure the metal blank to the hammerform base for forming the side panels.

This hammerform can be used to do two jobs. The radiused curved edge is used to form a metal edge on the wheel tub side. The flat straight edge can be used to bend a straight flange at the bottom of the side panels.

Tub Side Blank—Cut out a metal blank using your pattern. Clamp it between the hammerform base and top. Make sure it is tight. Use as many clamps as you can. Use a rawhide or

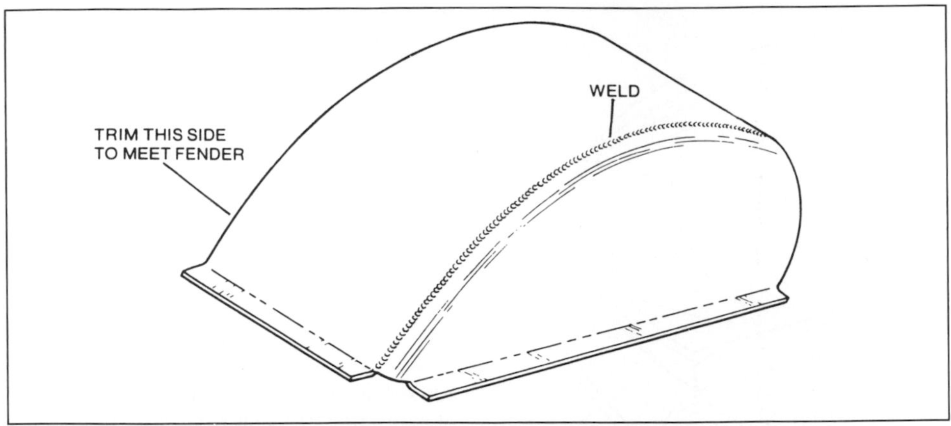

Finished hammerformed wheel tub should resemble sketch. Mounting flanges may or may not be straight, depending on floor panel they mate to.

wooden mallet to hammerform the edge. Work all around the curve gradually. Do not flog away in one spot. Turn the metal down over the radiused edge with repeated soft blows. It should form over the hammerform base very evenly. Be careful not to leave mallet marks in the metal.

If the side piece is to include a flange at the bottom, bend it up against the hammerform top. Make sure the hammerform top lines up with the intended bend line of the flange before you do any bending. Remove the wheel tub side from the form and fit it to the car at the frame. Clamp it in place and check for fit. If and when it fits properly, make the long strip that forms the wheel tub top.

Tub-Top Pattern—The long rectangle which makes the tub-top pattern can be fitted against the hammerformed side to double-check it for length. If you want it to include mounting flanges, add cardboard to the end of the pattern. Trim the pattern if it's too long. Now's the time to make any changes before going further.

Tub-Top Blank—Use the cardboard pattern to cut the metal blanks. If flanges are used, transfer their bend lines. Roll the blank to form a curve. Keep checking the blank against the hammerform base. When the strip is rolled close to the desired shape, check it in the car. Fit it over the side piece you clamped into place. Do any necessary adjusting to fit the pieces. The outer edge of the tub may require additional trimming to fit it to the inside of the quarter-panel. When the pieces fit closely, remove them from the car.

Welding the Tub—Tack-weld the wheel tub pieces together at the work bench. Before final welding, recheck their fit in the car. If the fit is correct, complete the welding. Cleco the finished tubs in the car. You can then build around and add to them with the other panels. The finished tubs have a nice curved inner edge, giving them more of a professional appearance than those built without a hammerform. Although it took more time and effort, it was worth it, right?

REAR FIREWALL

Next comes the rear firewall. It attaches to the rear wheel tubs and forms a wide flange at the bottom for the rear floor. Although considerably less complex than a front firewall, the rear firewall is constructed very much like it and is no less important. It must fit equally as well: no gaps or bulges. Follow the same rules used to construct the front firewall. *Be sure to seal any gaps or openings in the firewall securely.* As with the front firewall, add a 1-inch flange at the bottom edge of the rear firewall. This flange will support the rear edge of the floorpan and help secure it better.

In many cases the rear firewall is the main defense between the driver and the fuel tank. It has to be strong and well-sealed and remain so in the event of a crash. Precautions are necessary.

SIDE PANELS

You might think the floor panels should come next in the normal sequence of fabricating and fitting interior panels. Not so. Imagine having the floor in place. You would have to work on top of it or remove it to fit the side panels. Make the side panels next and you'll save the floor from the abuse of working on it. Or you will save the time to remove and reinstall it. *Make the side panels next.*

Side panels aren't as complicated as other interior panels. Virtually all side panels are flat, with the possible exception of a rolled edge at the top. Regardless, a good cardboard pattern is required to make a good piece.

The interior of a car is usually symmetrical like its exterior. The pattern for the side panel on one side can be used for the opposite panel. For example, the rear inside quarter-panel pattern can be used for both right and left sides. Door panels may need their own patterns because of different roll cage configurations in the right and left door openings.

Develop patterns for the side panels, transfer these patterns to sheet metal and cut out the blanks. Fit the panels one at a time.

Side Panel Beading—There is a simple way to bead side panels. I roll the bead a fixed distance from the panel edge so it follows the outline of the panel. The result stiffens the panel, particularly at its edge. The bead also gives emphasis to the shape of the panel. A nicely-shaped panel is enhanced. It makes your interior reflect the care you took to build it.

Forming—Except for mounting flanges, use large-radius bends rather than tight folds when forming side panels. This particularly applies to the top edges of door panels. It also applies to dashboards. Large radius bends give more of a professional look.

As you build each side panel, Cleco it in place. Each panel must fit to the other as well as the car. Don't install anything permanently. When all the side panels are Clecoed in place, go on to the floor.

FLOOR

Fabricating a floor is a complex and large job. It not only involves large panels, it also includes the transmission cover and the driveshaft tunnel.

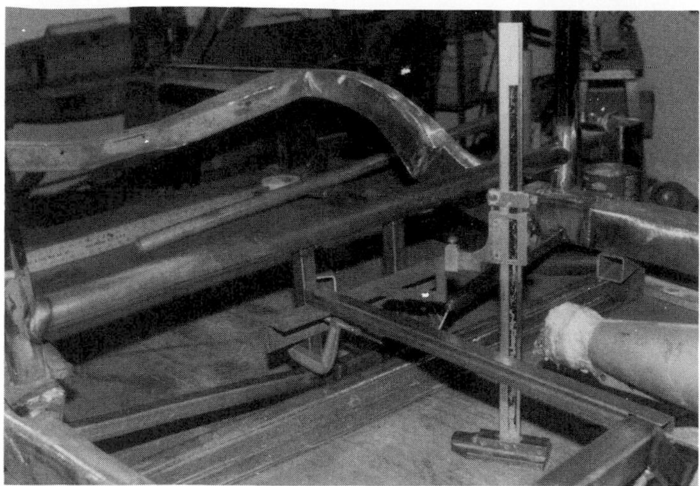

Adding tubes to existing structure makes a good floor foundation. Driver's side floor had to be very strong for seat mounts.

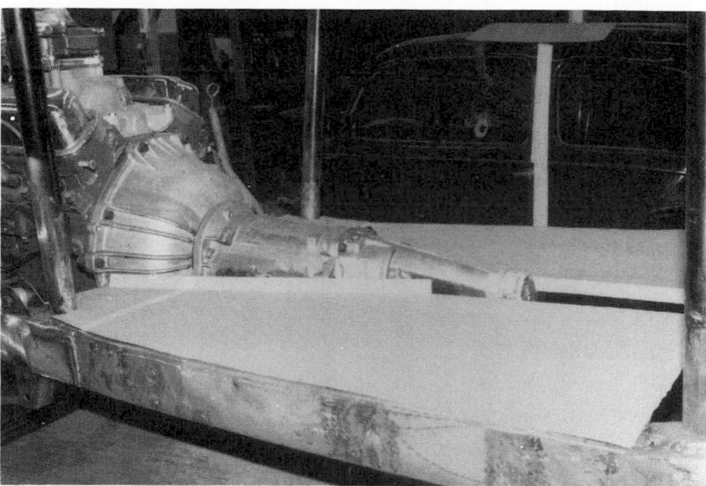

Floor pattern development is done after engine, transmission and all substructure is in place. Flange for transmission cover is part of this pattern.

Electric shears make quick work of cutting floor blank. Material is 0.040-in. mild steel. Foot area has already been beaded.

Some bends can't be made in a brake. Out come the C-clamps, angle iron and slapper. Presto, handformed bends.

These two areas require special attention to get right.

If a *live* axle is used—one that moves up and down with the wheels—the driveshaft tunnel *must* have enough clearance at the rear for vertical driveshaft travel. The transmission cover must include provisions for the shifter and the shifter boot. Therefore the shifter must be installed on the transmission. If the transmission cover fits closely around the transmission, the shifter *housing* will be a cube-like addition to the cover.

Floor Support—A floor should be strong and sturdy. It should have extra bracing immediately under the driver's feet. The floor will rest on flanges at the bottom edges of the front and rear firewalls. It will be supported at the sides of the car by one of several means. It may be bent up and riveted to the side panels. It might rest directly on top of an existing doorsill. The floor might even rest on angle brackets you welded to the rocker panel boxes. Or the floor may be attached directly to the frame rail top or side. There are many possibilities.

Use a felt-tip marker to indicate where trimming is necessary. Take your time fitting panels.

The floor is an excellent place to use beading. It provides a great practical benefit by increasing the strength of the flat floorpan. This is done without increasing its weight by one ounce. Although beading makes a panel more attractive, appearance is only a side benefit. A floor must be structurally sound to support the driver.

Floor Pattern—Right and left patterns are used to make floor panels. Start with the right side so you won't have to work around the seat mounts (if it's a left-drive single seater), and the pattern can be the basis for making

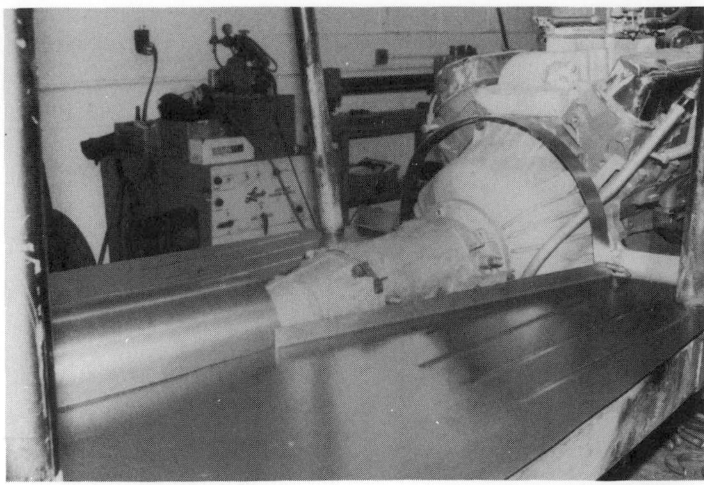

The Clecoed hoop in place is a "target." Cone-shaped transmission cover will be fitted to the hoop and driveshaft cover.

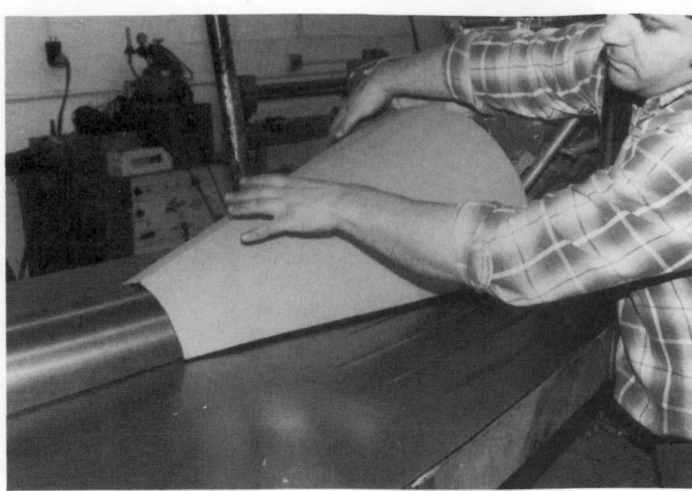

I'm developing the transmission cover pattern here.

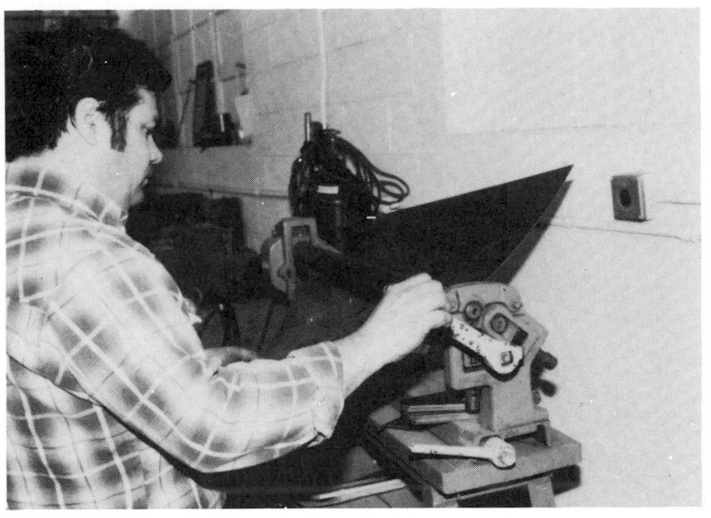

Rolling transmission cover requires that the roller be adjusted for more pressure at the small end to curve the blank into a cone.

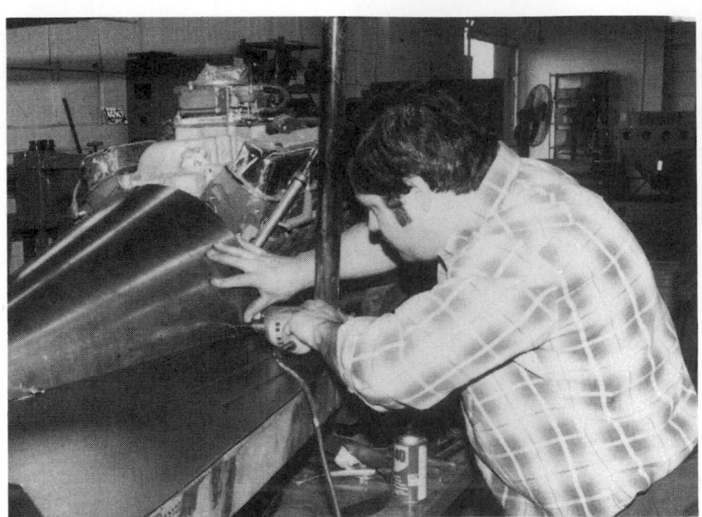

Fit transmission cover with care.

the left side pattern.

Sometimes, it's possible to fabricate the floor from a single sheet of metal because it doesn't have to be split its full length to clear the transmission or driveshaft. If this is the case, start with a piece of sheet metal at least as wide as the maximum interior width of the car. Find its center line and mark it so you'll have a reference when transferring right and left patterns to the blank.

After the right side pattern is developed and the blank is marked, flop the pattern over and check it for left side fit. You'll have to modify it to clear seat mounts and seat belt anchors. Adjustments may also have to be made at the engine, bellhousing, transmission and pedal box. Also allow for overall-width differences.

There must be flanges for mounting the transmission cover and driveshaft tunnel. These flanges can be added to the floor or the cover and tunnel. I recommend that you add flanges to the floor. They will provide additional stiffness to the floor. Add about 1-1/2-inches to the full length of the inside edge of each floorpan. Either way, flanges are needed. Your choice is *which* pieces get flanges and *how* the panels will interlock. There may also be flanges at other edges, depending on the design of your floorpan.

Once you have the floor blank cut out, bead the foot area first, then bend the flanges and do any other forming operations. Fit the panel to the car. Make changes to ensure the floorpan fits perfectly. When you're satisfied with the floorpan fit, Cleco the panel in place so you can develop the transmission cover and driveshaft tunnel.

Driveshaft Tunnel—Start with the driveshaft tunnel. It is relatively easy to build and you'll work to it with the transmission cover. You have a choice of two basic shapes; square or round. A square driveshaft tunnel has three flat surfaces: the top and both sides. The second shape, and the one I prefer, is round. It has one large curved, or rolled surface, that may blend into flat sides if the tunnel is deep. I like a curved tunnel because it is stronger than a square tunnel and I believe it looks more professional.

Checking transmission cover fit after Clecoing it in place. It fits well along the floor and is tight against the driveshaft tunnel.

Box-like addition to transmission cover accommodates shifter and boot.

Allow for maximum vertical driveshaft travel in your driveshaft tunnel pattern. Give the driveshaft 1-inch of clearance at its vertical travel limit. Simulate the cross-section of a round cover by looping a tape measure over the driveshaft from one floorpan half to the other. Make a 1- or 2-inch-wide metal test strip. Loop it over the driveshaft like you did with the tape, or put two breaks in the strip that are spaced the width of the top of the tunnel. It will be in the shape of a round or a square, depending on which tunnel shape you've chosen.

After you've established the cross-section of the tunnel, determine its length. It will run from the transmission tailshaft housing to the rear firewall or rear floor kickup. Make a cardboard pattern and check it for fit. Indicate flanges on the pattern if you didn't add flanges to the inner edges of the floorpan halves. Transfer the pattern to metal, cut out the blank, form it and Cleco the tunnel in place. Now make the last panel that will enclose the interior—the transmission cover.

Transmission Cover—Make a test strip like that pictured. Loop it over the transmission or bellhousing and provide at least a 2-inch clearance for the components underneath. Blend this section into the front of the driveshaft tunnel. If the driveshaft tunnel is round, the transmission cover will be cone-shaped. Using this test strip and the one for the driveshaft tunnel, develop a cardboard pattern. Don't forget the hole for the shifter. Make a metal blank, cut it out and roll the cone in a roller. Cleco the transmission cover in place.

Adding a *shifter box* to the transmission cover will complete the interior fabricating job. Determine the shape of the top of the box by matching it to the bottom of the shifter boot. The size of the hole in the top of the box is determined by how tightly the cover fits around the transmission and the *throw* of the shifter, both back-and-forth and from side-to-side. Build a cardboard box and tape it to the tunnel to check it for fit. When it's right, convert the box to metal. Bend the blank into a box using the cardboard pattern as your guide. Weld the box over the hole in the tunnel.

FASTENING PANELS

Unless you want them anodized, all you need to do with all those interior panels that are Clecoed in place is to secure them with permanent fasteners. Several methods can be used. How you fasten a panel depends on the raw material used and how many times you plan on removing it.

There is always some question as to how to fasten interior sheet metal panels. Several methods are available. *Quick-release* fasteners or rivets are commonly used. Stock-car steel interiors are sometimes MIG or TIG welded in place. You cannot weld an aluminum panel unless the mating parts are aluminum. You need to know something about all of the fastening methods available.

Screws—*Don't use sheet metal screws for fastening car interiors.* They have the undesirable habit of loosening and backing out of their holes. And there's that nasty sharp point that sticks through the panel. If you've ever worked on a car that was full of sheet metal screws, then you probably have the scars to prove it. Avoid them! There are better ways of fastening panels.

Once in a while I use *small machine screws*—10-32 to be exact—to retain certain panels. I always use *nylon-*

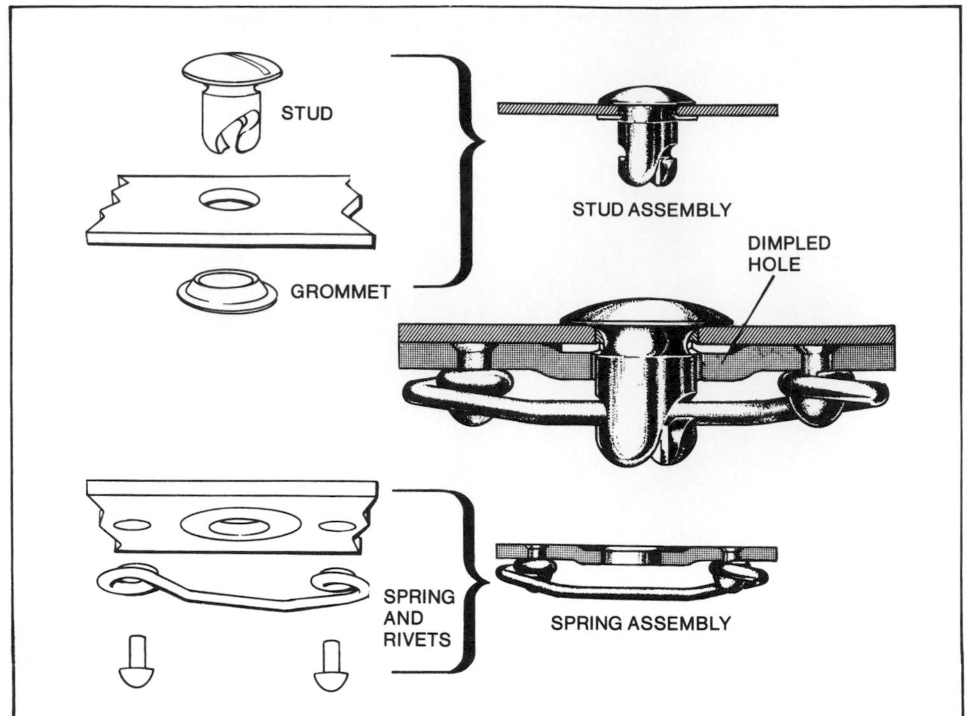

Details of AJ-series Dzus fastener installation.

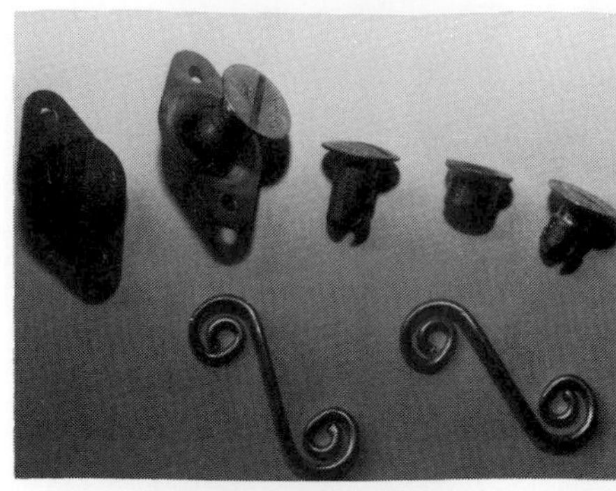

Dzus fasteners are worth their weight in gold! They hold panels securely and allow quick and easy removal. EHF self-ejecting fasteners at left. AJ series at right.

insert locknuts with them. This way the screws and nuts don't vibrate loose. Use these if you need to use screws for a particular panel. Don't use sheet metal screws!

Pop Rivets—Pop rivets are good for securing interior panels if done right: Use the right *type* and *number* of rivets. Use steel or stainless steel pop rivets. Unlike everyday aluminum rivets, they *stay* tight. Don't skimp on how many you use. The more rivets you use, the stronger and better sealed the interior will be. Don't space them more than 1-1/2-inch apart on long seams. Choose the correct grip length for the metal thickness you are joining. Refer to Chapter 8 on *Riveting* for determining rivet grip length. A rivet that's too long will not clamp panels tightly enough. Too short, and it won't hold at all. Be careful. One drawback to remember with rivets is that they are semi-permanent. The rivets would have to be drilled out to remove the panel, then new rivets need to go back in to secure it again. If your application will warrant frequent removal of these panels, such as with some racing cars, then you might want to consider an alternate fastening method.

Quick-Release Fasteners—
Unlike a screw that must be rotated several times to remove it, or a pop rivet that must be drilled out, a *quick-release* fastener releases when it is rotated just 90-degrees or 1/4-turn. This allows a panel to be removed very quickly. It can also be installed with the same relative speed. As a bonus, most quick-release fasteners stay with their panels when removed, so there is no danger of losing them. There are two well-known quick-release fasteners. *Dzus* or *Southco* fasteners are most commonly used in panels that may need to be removed quickly, frequently or repeatedly.

Two popular Dzus fasteners are in use. The EHF-series Dzus fasteners are *self-ejecting*. One half is mounted on a plate riveted to the panel it retains. This assembly includes the plate, a spring and the fastener. The fastener projects through holes in the metal panels the Dzus joins. It hooks over a *lock-wire* or spring, that is riveted to the other panel. When rotated 1/4-turn counterclockwise, the fastener unhooks from the lock-wire and the spring pops the fastener out, releasing the panel quickly.

The AJ-series Dzus fastener works much the same way. But it doesn't pop up automatically when released. The AJ fastener is held to the panel by an aluminum grommet. It is a smaller, more compact fastener than the EHF series. The AJ series includes flat and domed-head fasteners. Both EHF and AJ series fasteners require *dimpling* the panel to be installed.

Dimpling—*Dimpling*, similar to *belling*, starts with a drilled hole. The edge of the hole is then *dimpled*, or bent down into a specific shape before installing the lock wire. This allows the mating panels to fit against each other tightly.

Dzus fasteners are available in different lengths for joining different thicknesses of metal. They are available directly from the manufacturer. They can also be ordered from Earl's Supply, Russell Performance Products or Moroso. These companies also offer dimpling kits. Suppliers are usually helpful in suggesting the type or length of fastener if you'll explain what metals or parts are being joined.

Welding—Steel stock car interiors are sometimes welded in place. This makes an extremely strong, well-sealed interior. A welded interior even adds strength without adding much weight. The welded joints are well-sealed and nearly impenetrable under most circumstances—including accidents. Of all the fastening methods, a welded interior is the safest for the driver, offering the most protection.

HOW METAL IS LISTED

Typical Aluminum Listing:

MATERIAL	ALLOY	TEMPER	WIDTH X LENGTH X THICKNESS	NO. OF SHEETS	COST
Aluminum	3003	H-14	0.625" X 4' X 8'	1	$XX.XX

Technically, sheet aluminum and sheet steel are sold by weight. The cost per lb. fluctuates. The salesman may need to know the weight of an order to give you the price.

Typical Steel Sheet Listing:

MFRS. STANDARD GAGE AND SIZE	ESTIMATED WEIGHT PER SHEET	COST
18 ga. (0.0478")	48.0 lb.	Ask salesman
Wt. per Sq. Ft. 2.00	48.0 lb.	Ask salesman
36 x 96	60.0 lb.	Ask salesman
36 x 120	60.0 lb.	Ask salesman
48 x 120	80.0 lb.	Ask salesman

HAMMERFORM COMPATIBILITY CHART

Material to Be Formed	Thickness	Hammerform Material	Corking Tool	Comments
1100 series Aluminum	0.050" 0.063" 0.080"	Hardwood (up to 5 parts) 6061 T6 or T4 aluminum (more than 5 parts)	Hardwood	Soft; hammerforms easily
3003-H14 Aluminum	0.050" 0.063" 0.800"	Hardwood (1 part) 6061 T6 aluminum (more than 2 parts)	Hardwood	Harder than 1100 series aluminum; very popular to use in hammerforming
6061 T4 or T6 Aluminum	0.050" 0.063" 0.080"	Generally used as the hammerform material; not hammerformed itself		Not good for hammerforming; too hard; may crack
1020 Cold-Rolled Steel	18 ga. 20 ga.	Hardwood (1 part) 6061-T6 aluminum (up to 10 parts)	aluminum or hardwood	Good for hammerformed brackets; good for small steel body parts
1020 AK or SK Steel	20 ga. 18 ga. 16 ga. 14 ga.	6061 T6 aluminum (up to 10 parts) 1020 CR steel (over 10 parts)	aluminum or hardwood	Good for hammerformed brackets, but more expensive than 1020 CR steel.
4130 Chrome-Moly Steel	20 ga. 18 ga. 16 ga.	1020 mild steel (any number of parts)	aluminum or steel	Good for suspension parts and other brackets requiring strength; expensive

U.S. STANDARD SHEET METAL GAGES FOR STEEL

SPECIFY MATERIALS IN DECIMALS OF AN INCH WHEN ORDERING

Number of Gage	Thickness (Fraction of Inch)
3	0.2391
4	0.2242
5	0.2092
6	0.1943
7	0.1793
8	0.1644
9	0.1495
10	0.1345
11	0.1196
12	0.1046
13	0.0897
14	0.0747
15	0.0673
16	0.0598
17	0.0538
18	0.0478
19	0.0418
20	0.0359
21	0.0329
22	0.0299
23	0.0269
24	0.0239
25	0.0209
26	0.0179
27	0.0164
28	0.0149
29	0.0135
30	0.0120
31	0.0105
32	0.0097
33	0.0090
34	0.0082
35	0.0075
36	0.0067
37	0.0064
38	0.0060

METAL ALLOY QUICK REFERENCE GUIDE

Alloy & Temper	Common Sheet Size & Thickness	Welding/Bending/ Shaping Qualities	Common Uses	Comments
1100 Aluminum	4' x 10' 4' x 8' 0.040" 0.050" 0.060"	Excellent for all shaping, bending and welding	1100-H14 great for hand-formed bodies, hammerformed parts, or spun parts.	1100-0 is very soft, easy to shape. 1100-H-14 becomes very bright and shiny when shaped by English wheel.
6061-T6 Aluminum	4' x I0' 4' x 8' 0.040" 0.050" 0.063"	Bendable, but may crack if bent on tight radius. Weldable, but welds may tend to crack	Highly stressed structural parts like Indy-car tubs.	Strong but brittle. Best joining method: rivets or bonding.
6061-T4 Aluminum	4' x 10' sheets 4' x 8' sheets 0.040" 0.050" 0.063"	Bendable, but cracks easily. Weldable, but welds may crack easily.	Ideal for monocoque construction. Good for car wings.	Good strength. Some brittleness. Riveted and bonded for joining or installing.
3003-H14 Aluminum	4' x 10' sheets 4' x 8' sheets 0.040" 0.050" 0.063" 0.080"	Excellent for bending, good for shaping, very good for welding by heliarc or oxylacetylene.	Body panels, aluminum interiors, baffles, heat shields, surge tanks, air scoops. Many other applications.	Very attractive, anodizes well, easily bent or shaped, polishes well, can be annealed to restore workability if it work-hardens.
1020 Cold-Rolled Steel	4' x 10' sheets 4' x 8' sheets 0.0299" 0.0478" 0.1196"	Excellent for bending. Good for shaping. Excellent for welding.	Roll bar mounts, fuel cell containers, fuel tanks. Hammerformed brackets. Bodywork. Floorpans. Many other applications.	Low cost. Excellent availability. Strong. Cuts easily.
1020 AK or SK Steel	4' x '10 sheets 4' x 8' sheets 0.0299" 0.0359" 0.0478"	Good for forming, shaping, or bending. Excellent for welding.	General bodywork. Hammerformed brackets.	Very easily formed. Limited availability. Expensive.
4130 Chrome-Moly Steel	4' x I0' sheets 4' x 8' sheets 0.0299" 0.0359" 0.0478" 0.0598" 0.1196"	Good for bending in annealed state. Not recommended for shaping components. Good for welding.	Fabricated suspension arms, other suspension roll bars, roll bar gussets, anti-sway bars, brackets, mounting plates for roll bars.	Stronger than 1020 CR steel of same gage. Limited availability. Very expensive. Dulls cutting tools easily and quickly.

GLOSSARY

Align—To arrange in correct position; to place in a straight line.

AK Steel—A steel alloy which has been "killed" with aluminum in the molten stage to refine its grain structure; a steel alloy with good ductility.

Alloy—A blending of metals; homogeneous combination of metals in which the atoms of one replace or occupy positions between atoms of the other.

Aluminum Alloy—A range of metals combining aluminum with copper, magnesium, zinc, nickel and/or iron.

Anneal—With aluminum, a controlled process of heating metal to 640-degrees F. (338C) and cooling rapidly to 450-degrees F. (232C) until recrystalization occurs, in order to soften the metal; in steel, a heating and cooling operation of steel in the solid state usually requiring slow gradual cooling.

Anodize—To coat a metallic surface electrolytically with a protective or colorative oxide.

Anvil—The lower wheel of the English wheeling machine.

Baffle—An object designed to block or damp the movement of heat, air or liquid.

Base Station—In metal-work, the bottom section of a station buck that acts as a reference point.

Bead—In welding, a narrow half-round pattern where metal has been joined by heating; in metal-working, a decorative or structural half-round channel formed into the metal in a continuous line.

Beader—A piece of metal-working equipment used to indent, step or groove metal for structural or decorative purposes.

Bed Width—The overall width of a whole press brake machine.

Bell—To turn down the edge of a hole in sheet metal with a smooth curve.

Bend—To cause something to assume a curved or angular shape; something which has been bent; a curve or angle.

Bend Line—The line indicating the placement of a bend; the line upon which metal is bent.

Bending Capacity—Thickness and width of metal a sheet metal brake can bend.

Bending Sequence—The order in which several different bends are formed so that each successive bend is not blocked by a previous bend.

Blade—One arm of a squaring tool; cutting edge of a shearing machine.

Blank—A cutout metal piece prepared for fabrication.

Blanking—Measuring out the general overall shape of a piece of metal before beginning to shape it.

Body Filler—Lead or plastic used to fill in low spots on a body panel during restoration or repair.

Body Seam Filler—Special paint-like sealer used to completely seal a seam not fully welded.

Brake—To bend a metal part to the proper radius and angle on a brake machine or by hand; the machine which forms angles or folds in metal.

Bucking Bar—See Rivet Buck

Burnish—To make smooth or glossy as if by rubbing; polish, planish.

Butt-Weld—Welding two metal pieces together at their butted edges.

Center Line—In patterns and layout, a line indicating the exact center of a given part, from which other details of layout can be determined.

Chipboard—A lightweight flexible cardboard used to make patterns.

Chrome-Moly—A strong and expensive steel alloy, designated SAE 4130,, that is composed of 0.28-0.33 % chromium and 0.15-0.25% molybdenum.

Collector—A metal assembly in an exhaust manifold that collects exhaust gas flow from the primary exhaust tubes of an engine, funneling this flow into a single exhaust pipe.

Concave—Curved like the inside of a bowl; hollow.

Convex—Curved like the exterior surface of a bowl; curving or bulging outward.

Corking Tool—A 6- to 9-in. long tool of hardwood, aluminum or steel used to transfer the impact of a hammer to the metal blank held in a hammerform.

Cross-Section—A section formed by a plane cutting directly through an object.

Crown—A metal-work term used to define the overall height of a raised, curved area.

Deburr—To remove uneven or jagged edges from a metal piece.

Die—A part on a machine forming sheet metal, generally made of a metal harder than the metal it is used to form.

Dimple—To turn the edge of a hole under or down to accept installation of a fastener; to bell.

Direct Layout—Marking metal according to dimensions needed for a given piece without first making a pattern or template; used for making simple pieces.

Dolly—In metal-work, a tool used underneath metal being hammered from above; to shape metal using a dolly.

Ductility—Capability of a metal to be shaped or hammered without cracking or breaking; flexibility.

File Cut—The direction and texture of the rows of cutting teeth arranged across the surface of a file.

Fillet-Weld—Welding two metal pieces which join at an angle to each other; this weld usually fills the inner corner of the angle, sometimes the outer.

Finger—A part of a box finger brake which attaches to or detaches from the upper beam to facilitate bending box or pan shapes.

Fitting—A general term for any small part, such as plumbing, used in the structure of a component.

Fixture—A device used to hold a part being formed securely; a jig.

Flange—A protruding and angled edge used to strengthen or to attach one object to another.

Floorpan—A metal part between frame rails, under seats, upon which feet rest in the car.

Flux—Fusible paste or powder used to facilitate brazing steel with brass or bronze welding rod; also a fusible paste or powder used in gas welding aluminum or silver soldering. Flux may also be coated on a welding or brazing rod.

Fold—To bend, or brake; an angle formed in metal.

Folder—In metal-work, a name commonly used for a combination sheet metal brake.

Gage—Standard thickness of metal; measurement rule mounted on a metal cutting machine.

Graduation—On measuring devices, the marks dividing the whole at even intervals.

Hammerform—A wood, steel or aluminum shape used as a form to shape metal hammered over or into it; the act of shaping metal using such a form.

Heat Treatable—Metal alloys that can be hardened by heating; heat-treatable aluminum bears the letters H or T.

Heliarc—A kind of electric welding named for the original shielding gas, helium, and the electric arc which is the heat source; also known as TIG welding.

Hollow—In American metal-working usage, deeply indented or concave; a metal shaping technique used to form a depression in metal; sometimes referred to as "raising" in British metal-working.

Indent—In metal-work, to use a beader with an indenting die to form a groove on the bend line to facilitate bending a blank.

Jig—A tool or fixture holding a part upon which some operation is being performed.

Joggle—See Step.

Layout—In metal-work, full-size outline of parts marked on sheet metal; for designer, a full-scale drawing of parts to be fabricated or assembled.

Layout Dye—A liquid dye, also called *Dykem* or *Machinists' Blue*; used to coat a metal surface during layout so markings will be more visible; also used during metal finishing to make low spots more visible.

Leaf Brake—See Sheet Metal Brake.

Lightening Hole—A hole made in any metal part to reduce overall weight by removing metal.

Locator Clamp—In hammerforming, a clamp used to secure the blank to the hammerform.

Locator Ring—In hammerforming, a ring used to hold the metal blank in position over the hammerform bottom as it is formed.

Locating Pins—In hammerforming, pins used to keep the hammerform top and metal blank securely attached to hammerform bottom.

Mallet—A short handled kind of hammer, with a cylindrical head of wood, plastic or rawhide.

Mandrel—A metal core over which material may be shaped.

Metal Finish—To smooth, polish and make even by hammering, grinding, filing or rubbing.

Mild Steel—A low-carbon steel alloy containing 0.030-0.050% carbon.

Normalize—Low and medium carbon steel heated 50-100-degrees F.(10-38C) above critical temperature range and allowed to cool slowly in still air; used to refine the grain and to stress-relieve.

Panel Beating—An English automotive term used to mean hand-forming large automobile body panels.

Parallel—Straight or curved lines equidistant from one another at every point.

Pattern—A detailed paper, chipboard or metal form used to indicate all important details of a finished part or piece.

Penetration—The depth reached by an object or force; the degree to which heat penetrates metal during welding.

Perpendicular—Intersecting at right angles (90 degrees); at a right angle to either horizontal or vertical.

Pi—The relationship between the diameter and the circumference of a circle, approximately 3.1416.

Pilot Hole—Small hole drilled to guide final drilling, so the drill bit or chassis punch will be accurately placed.

Planish—To finish, make smooth or harden metal by pounding out irregularities; to smooth, toughen, flatten or polish metal by rolling or hammering; to burnish.

Plug Weld—Also known as a rosette weld, weld joins two overlaying metal pieces through a hole in the upper piece; weld fills the hole.

Pneumatic—Operated or powered by compressed air.

Push block—A wooden block used to guide metal into the teeth of a band saw blade, or into the sanding disc of a combination sander.

Radius Bend—In metal-work, a curved fold as opposed to a straight angle fold.

Raise—A method of stretching metal by hammering it into a depression such as a hollowed wood base or shot bag; sometimes used to mean hollowing.

Rams—Hydraulic cylinders on a press brake which control the upper die's movement.

Rendering—To represent in detail in an artistic form; a precise scale picture of a detailed construction.

Restoration—To restore something to its original shape, place or condition.

Rivet Buck—A hardened metal bar held against the rivet stem while setting rivets to facilitate formation of the shophead.

Rivet Head—The rounded end of a rivet, formed during manufacture.

Rivet Set—The part of a rivet gun which fits onto the rivet head and transfers power from rivet gun to rivet.

Rivet Stem—The metal rod of a rivet, formed during manufacture.

Roller—Also called slip roller or rolling machine, a metal-working machine which curves or rolls metal in a single plane.

Rosette Weld—See "Plug Weld".

Scale—In welding, a flaky oxide film of metal impurities sometimes formed on a metal, particularly iron alloys, when heated to high temperatures; in drawing, a method to represent the proportions of a measurement.

Semi-Circle—Half a circle as divided by the diameter.

Sheet Metal Brake—A piece of metal-working equipment used to bend metal.

Shop Head—The head of a solid rivet which is formed as the rivet is set.

Shot Bag—A leather bag filled with very fine lead shot and sewn shut over which metal is shaped; #9 bird shot is most desirable for metal-working.

Shrink—To draw together and reduce the amount of metal area; usually resulting in an increase in metal thickness

Simple Curve—A curve on one plane of a surface.

Skin—The smooth outer layer of sheet metal overlaying an inner structure.

SK Steel—A steel alloy which has been "killed" with silicon in the molten stage to refine its grain structure; a steel alloy with good ductility.

Slapper—Long, rectangular metal or wood tool used to curve, flatten or smooth a metal surface; sometimes a wooden slapper is covered with leather.

Slip Roller—A piece of metal-forming equipment used to form curves or tubular shapes, also commonly called a roller or roll.

Spring Back—The tendency of sheet metal or tubing to attempt to return to its original shape after bending or forming.

Station—Wooden piece to represent a cross-section of a metal construction at a given point.

Station Buck—Wooden form representing the shape of a metal part; a device for ensuring accuracy in metal construction.

Step—To offset, or raise or lower a metal surface by forcing it through a die; usually the metal is raised or lowered the amount of the metal thickness itself.

Stretch—To lengthen or widen metal by hammering or rolling; to increase the area of sheet metal by hammering or rolling, resulting in thinning as a secondary effect.

Stroke—(as in stroke gage)—the distance an upper die travels down into a lower die.

Structural Integrity—Structure which has the quality of being sound throughout.

Symmetrical—A shape identical in form and configuration on either side of a center line.

T-Dolly—A hand tool shaped like a capital T, with a flat stem and rounded dolly on top.

Tack-Weld—A welding method using small areas of weld spaced along a seam between adjoining surfaces; a temporary joining to prepare for final welding.

Tangent—In metal-work, the beginning or end of a curved bend.

Temper—Hardness of a metal that governs its strength and formability; to reheat metal after hardening to a temperature lower than critical range and then cooling.

Template—A pattern for laying out parts on a flat sheet of metal; a pattern showing bend lines, flanges or other special information about construction of the metal piece, including bending allowances.

Throat Depth—The distance between the operating part of a band saw, planishing hammer, beader or Kraftformer and the machine's housing or frame; the greatest distance possible to reach into the metal from the edge with one of these machines.

Tinning—In restoration or repair, a process of preparing the metal surface to accept lead as a filler.

Tinning Compound—Also called tinning paste, a material used as a flux in restoration or repair work to prepare the metal body surface to accept lead.

Tracking—Moving metal through an English wheel in one continuous pass; or the pattern of such passes.

Wheelhouse—The part of a car body which acts as an inner fender.

Witness Line—A reference mark for accurately fitting or joining two parts.

Work-Harden—An increase in the hardness and strength of metal as it is formed or worked; an undesirable over-hardening is sometimes encountered when metal is shaped.

SUPPLIER'S INDEX

CUSTOM METAL SHOPS:

BIZER ASSOCIATES, INC.
30701 Glenmuer
Farmington Hills, MI 48018

FAY BUTLER
P. 0. Box 106
Wheelwright, MA 01094

BRITALIA COACHWORKS
R R 1, P. 0. Box 8
West Creek, NJ 08092

COACH CRAFT, INC.
P. 0. Box 728
Mooresville, NC 28115

CUSTOM METAL SHAPING
1270 No. Lance La.
Anaheim, CA 92807

DAVIS CARS
17702 Metsler Lane
Huntington Beach, CA 29647

D.F. METALWORKS
7712 Talbert Ave.
Unit A
Huntington Beach, CA 92647

ENTECH METAL FABRICATION
900 Chicago Rd.
Troy, MI 48083

HEGMAN'S SPECIALTY CARS
18109 Mt. Washington
Fountain Valley, CA 92708

HOT RODS BY BOYD, INC.
8372 Monroe
Stanton, CA 90680

JOCKO & CO.
73738 Desert Trail
Twentynine Palms, CA 92277

KLEEVS COMPANY
4526 Griswold Rd.
Port Huron, MI 48060

LIGHT ENTERPRISES
Rt. 6, P.0. Box 340
Hagerstown, MD 21740

M&G VINTAGE CAR RESTORATIONS
3570 Wolfendale Rd.
Mississauga, Ontario,
Canada L5C 2V6

M&L AUTO SPECIALISTS
1212 Washington St.
Manitowoc, WI 54220

SCOTT'S HAMMER WORKS
19231 E. San Jose Ave.
City of Industry, CA 91748

MEL SWAIN
4392 Blg. Two, Unit D, Brooks St.
Montclair, CA

DICK TROUTMAN
3198 L Airport Loop Dr.
Costa Mesa, CA 92626

UNIQUE METAL PRODUCTS
8745 Magnolia
Santee, CA 92071

DENNIS WEBB DESIGNS
831 So. Lime St.
Anaheim, CA 92805

HAND TOOL SUPPLIERS:

THE EASTWOOD CO.
580 Lancaster Ave.
Malvern, PA 19355

T. N. COWAN
P. 0. Box 900
Alvarado, TX 76009

MALCO PRODUCTS, INC.
8710 Science Center Dr.
Minneapolis, MN 55428

MILLERS FALLS
57 Wells St.
Greenfield, MA 01301

NIAGARA MACHINE & TOOL WORKS
P. 0. Box 475, 633 Northland Ave.
Buffalo, NY 14240

PEXTO
Peck, Stow & Wilcox, Co.
Southington, CT 06489

ROCKWELL DELTA
400 Lexington Ave.
Pittsburgh, PA 15208

ROPER WHITNEY, INC.
2833 Huffman Blvd.
Rockford, IL 61101

SEARS ROEBUCK & CO.
Chicago, IL 60684
(check local area)

SNAP-ON TOOLS CORP.
Kenosha, WI 53140

L.S. STARRETT CO.
Athol, MA 01331

STANLEY PROTO TOOLS
3029 Bankers Industrial Dr.
Atlanta, GA 30362

U.S. INDUSTRIAL TOOL & SUPPLY
15101 Cleat St.
Plymouth, MI 48170

LARGE METAL SHAPING EQUIPMENT:

W. ECKOLD, A.G.
Ch 7202 Trimmis-Station
Switzerland

NIAGARA MACHINE & TOOL WORKS
(see address listed)

PEXTO
(see address listed)

ROCKWELL DELTA
400 Lexington Ave.
Pittsburgh, PA 15208

ROPER WHITNEY, INC.
(see address listed)

ROTEX PUNCH CO.
2350 Alvarado St.
San Leandro, CA 94577

U.S. INDUSTRIAL TOOL & SUPPLY
(see address listed)

WELLS MANUFACTURING CORP.
407 Jefferson St.
Three Rivers, MI 49093

WILLIAMS LOW BUCK TOOLS
4175 California Ave.
Norco, CA 91760

POWER TOOLS:

BLACK & DECKER U.S., INC.
Towson, MD 21204

ROBERT BOSCH CORP.
2800 South 25th Ave.
Broadview, IL 60153

THE EASTWOOD CO.
(see address listed)

MAKITA U.S.A., INC.
12930 E. Alondra Blvd.
Cerritos, CA 90701

MILLERS FALLS
(see address listed)

MILWAUKEE ELECTRIC TOOL CORP.
13135 West Lisbon Rd.
Brookfield, WI 53005

SEARS ROEBUCK & CO.
(check local area)

U.S. GENERAL
100 Commercial St.
Plainview, NY 11803

SAFETY PRODUCTS:

NORTON SAFETY PRODUCTS
2000 Plainfield Pike
Granston, RI 02920

SEARS ROEBUCK & CO.
(check local area)

ABLES

METRIC/CUSTOMARY-UNIT EQUIVALENTS

Multiply:		by:		to get:	Multiply:		by:		to get:
LINEAR									
inches	X	25.4	=	millimeters(mm)		X	0.03937	=	inches
feet	X	0.3048	=	meters (m)		X	3.281	=	feet
miles	X	1.6093	=	kilometers (km)		X	0.6214	=	miles
AREA									
inches^2	X	645.16	=	millimeters^2(mm^2)		X	0.00155	=	inches^2
feet^2	X	0.0929	=	meters^2(m^2)		X	10.764	=	feet^2
VOLUME									
inches^3	X	16387	=	millimeters^3(mm^3)		X	0.000061	=	inches^3
inches^3	X	0.01639	=	liters (l)		X	61.024	=	inches^3
quarts	X	0.94635	=	liters (l)		X	1.0567	=	quarts
gallons	X	3.7854	=	liters (l)		X	0.2642	=	gallons
feet^3	X	28.317	=	liters (l)		X	0.03531	=	feet^3
feet^3	X	0.02832	=	meters^3(m^3)		X	35.315	=	feet^3
MASS									
pounds (av)	X	0.4536	=	kilograms (kg)		X	2.2046	=	pounds (av)
FORCE									
pounds—f(av)	X	4.448	=	newtons (N)		X	0.2248	=	pounds—f(av)
kilograms—f	X	9.807	=	newtons (N)		X	0.10197	=	kilograms—f

TEMPERATURE

Degrees Celsius (C) = 0.556 (F - 32)

Degrees Farenheit (F) = (1.8C) + 32 × 40

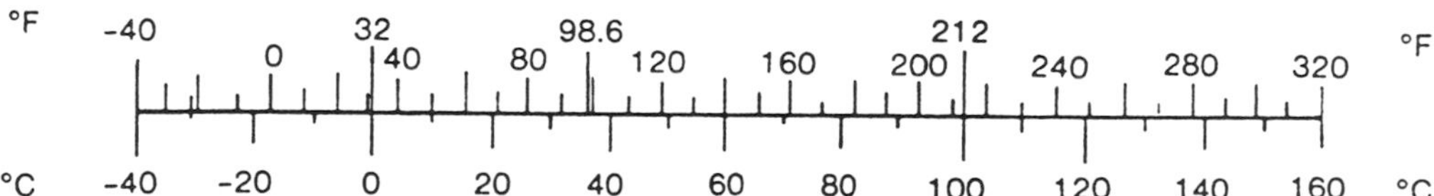

ACCELERATION

feet/sec^2	X	0.3048	= meters/sec^2(m/s^2)	X	3.281	=	feet/sec^2
inches/sec^2	X	0.0254	= meters/sec^2(m/s^2)	X	39.37	=	inches/sec^2

ENERGY OR WORK (Watt-second = joule = newton-meter)

foot-pounds	X	1.3558	= joules (J)	X	0.7376	=	foot-pounds
calories	X	4.187	= joules (J)	X	0.2388	=	calories
Btu	X	1055	= joules (J)	X	0.000948	=	Btu
watt-hours	X	3600	= joules (J)	X	0.0002778	=	watt-hours
kilowatt-hrs	X	3.600	= megajoules (MJ)	X	0.2778	=	kilowatt-hrs

FUEL ECONOMY & FUEL CONSUMPTION

miles/gal	X	0.42514	= kilometers/liter(km/l)	X	2.3522	=	miles/gal

Note:
235.2/(mi/gal) = liters/100km
235.2/(liters/100km) = mi/gal

PRESSURE OR STRESS

inches Hg (60F)	X	3.377	= kilopascals (kPa)	X	0.2961	=	inches Hg
pounds/sq in.	X	6.895	= kilopascals (kPa)	X	0.145	=	pounds/sq in

POWER

horsepower	X	0.746	= kilowatts (kW)	X	1.34	=	horsepower

TORQUE

pound-inches	X	0.11298	= newton-meters (N-m)	X	8.851	=	pound-inches
pound-feet	X	1.3558	= newton-meters (N-m)	X	0.7376	=	pound-feet
pound-inches	X	0.0115	= kilogram-meters (Kg-M)	X	87	=	pound-inches
pound-feet	X	0.138	= kilogram-meters (Kg-M)	X	7.25	=	pound-feet

VELOCITY

miles/hour	X	1.6093	= kilometers/hour(km/h)	X	0.6214	=	miles/hour
kilometers/hr	X	0.27778	= meters/sec (m/s)	X	3.600	=	kilometers/hr

COMMON METRIC PREFIXES

mega	(M)	=	1,000,000	or	10^6	centi	(c)	=	0.01	or	$10^{\times 2}$
kilo	(k)	=	1,000	or	10^3	milli	(m)	=	0.001	or	$10^{\times 3}$
hecto	(h)	=	100	or	10^2	micro	(μ)	=	0.000,001	or	$10^{\times 6}$

INDEX

A
Air Planishing Hammers 69, 70
  Pneumatic 70
  Design 69
  Throat Depth 70
  Usage 70
Air Scoop 60
Aluminum 42-44
  1100 Series 43
  2024 Series 44
  3003-H14 Series 44
  6061 Series 44
  Coding 42
  Cold-working 44
  Hardness 44
  Work-hardening 44
Annealing 57, 44

B
Band Saws 35
  Blade Guide 36
  Blade Welder 36
  Work Table 36
  Measurements 35
  Forming Blades 36
Beaders 33
  Joggling 34
  Stepping 34
  Indenting 34
  Bending 63
  Dies 63
Benches 12
  Layout 12
  Work 12
Body Fillers 107
Brakes 12,27-32
  Box Finger 30
  Combination 30, 61
  Radius Bends 30, 61
  Radius-Forming Dies 30
  Straight Bending 30
  Components 28
  Bending Capacities 28
  Lower Beam 28
  Upper Beam 28
  Making Bends 30, 61

C
Catalogs 41
  Ordering Sheet Metal 42
Chassis Punches 25
Chrome-Moly 46, 47
  4130 46
  Annealed 47
  Normalized 47
  Stress-Relieved 47
  Tensile Strength 46
Corking Tools 82
  Aluminum 82
  Hardwood 82
  Length 82
Cutting Tools 15

D
Dimpling 124
Dividers 21
Dollies 18
Dollying 101, 102
Drills 16
Driveshaft Tunnel 122

E
English Wheel 67
  Anvils 67
  Design 68
  History 67
  Tracking 69
  Use 68

F
Fasteners 123
  For Panels 123
  Dzus 124
  Lock Wire 124
  Pop Rivets 123
  Quick Release Fasteners 123,124
  Screws 123
  Self-Ejecting 124
  Southco 124
Files 16, 17, 103, 104 110, 111
  Cut 17
  Taper 17
  Length 17
  Vixen 103
Final Planishing 102, 59
Firewalls 116-120
  Front 116
  Beading 118
  Fitting 118
  Flanging 118
  Installation 118
  Pattern Details 117
  Pedal box 117
  Stepping 118
  Rear 120
Fitting Body Panels 105
Flanges 62
Flaring Tools 25
  Lightening holes 25
Floors 120
  Pattern 121
  Support 121
  Live Axle 121
Floorpan Restoration 106
  Body Seam Filler 107
  Plug Weld 107
  Spot Weld 107
  Flanges 62
  T-Dolly 64

G
Glass Pliers 24

H
Hammerform Construction 78
  Blank Pattern 79
  Metal Blank 79
  Securing 80
Hammerform Design 77
Hammerforming 75, 76, 77, 119
  Concave Shape 76
  Convex Shape 76
  Working With 81
Hammering 58, 59 82
Hammers 17,18
  Dinging 18
  Pick 18
  Shallow-domed Face 18
  Mallets 18
Helper Tools 22
  C-Clamps 23
  Clecos 23
  Cleco Pliers 23
  Vise 23
  Vise Grips 23
  Vise Jaw Covers 23

I
Interiors 114-124
  Drag Racing 113
  Street Machines 113
  Road Racing 114
  Circle Track 114
  Driver Protection 115

K
Kraftformers 66, 67

L
Layout 49, 52, 61
  Rules 52
Layout Tools 18
Lead Fillers 107-111
  Prep 108
  Tinning 108
  Application 109
  Filing 110
Leaf Brake 27

M
Metal Finishing 59, 100
  Body & File Sander 103
  Dollying 101
  Final planishing 102
  Vixen files 103
  Low spots 104, 105
  High spots 104
  Shrinking With Heat 104
  Weld Finishing 102
  Shrinking & Leveling Disk 105
Metal Shaping 55
  Shot Bags 56
  Stretching vs. Shrinking 56
  Cold-Shrinking 56
  Hot-Shrinking 56

N
Notchers 25

P
Patterns 49-53, 57, 61
  Marking 51, 52
  Information 52
  Fitting 52
  Development 52
  Materials 50
  Chipboard 50
  Metal 52
Placing Equipment 11
  Shear 12
  Sheet Metal Brakes 12
Plastic Fillers 111
  Application 111
  Cross-Hatch Pattern 111
  Filing 111
Pop-Rivet Gun 23
Power Hammers 71, 72
  Design 71
  History 71
  Dies 72
  Cross Die 73
  Curved Die 73
  High Crown 73
  Low Crown 73
  Shrinking Dies 73
Press Brake 37
  Bed Width 37
  Bending Dies 37
  Rams 37
  Ram Speed 37
  Tonnage 37
  Using Press Brakes 38
  Back Gage 39
  Stroke Gage 39
Punches 21
  Automatic Center 22
  Center 21, 22
  Prick 21, 22
  Hand-lever 22
  Transfer 22

R
Reproductions 99
Research 99
Restoration 97
  Original Sheet Metal 99
  Reverse Patterns 98
  Patterns & Information 98
Restoration Shops 99
Rivet Bucks 92
Riveting 89-93
  Advantages 89
  Sheer Strength 89
  Preparing Rivet Holes 93
  Pneumatic Rivet Guns 91
Rivet Types 89
  Solid 91
  Blind 90
  Pop 90
  Stem 89
  Shop head 89, 91
  Head 90
  Universal 92
  Round 92
  Flat 92
Rotex Punch 37
  Turret 37
  Punch Throat 37
  Usage 37
Rollers 33

S
Sanders 36
  Combination 36
  Disc 36
  Dust Collectors 36
  Sizes 36
Scribers 21, 31
Shaping Equipment 64
Shaping Tools 17
Shears 12
  Hand 16
Sheet Metal Interiors 113
  3003-H14 113
  Rules 115
  Circle Track 114
  Drag Cars 113
  Driver Protection 115
  Installation 116
  Road Racing Cars 110
  Street Machines 113
Sheet Metal Racks l3
Sheet Metal Rollers 32
Sheet Metal Shears 32
  Back Gage 32
  Foot Shear 32
  Locator Clamp 32
  Power Squaring 32
  Side Gage 32
  Using Shears 32
Shifter Box 123
Shop Safety 13
Shrinkers & Stretchers 34, 35
  Throat Depth 35
Shot Bag 56
Shrinking Steel with Heat 104
Slappers 26
  Shot Bag 26
Snips 15
  Aviation 15
  Tinner's 15
Squares 18,20
  Carpenter's 18
  T-Square 19
  Panel 19
  Combination 19
  Using squares 19
Station Bucks 65
  Stations 65
  Tacking Jig 65
  Construction 65
  Horizontal Base Station 65
Steel 45
  Coding 45
  Carbon Content 45
  1020 Cold-Rolled 45
  AK & SK steel 45
  Deep Draw Qualities 45
  Killing 45
Steel Rules 20

T
T-Dollies 25, 26, 64
Tack-Welding 100
Tape Measures 20
Tear-Drop Shaped Bowl 56,58
Templates 57
Tinning 108
  Compound 108, 109
  Acid 108, 109
  Application 109
  Working 110
  Beeswax 110
  Filing 110
  Cleaning 110
  Metal Prep 110
  Solder 111
  30/70 Lead 111
Transmission Covers 122, 123

W
Weld Finishing 102
  Excess Weld 102
Welding 124
Welding the Tub 120
Wheelhouses or Tubs
  Measurements 118
  Expansion Factor 118
  Shape 119
  Tub-Side Blank 119
  Tub-Top Blank 120
  Tub-Top Pattern 120

OTHER BOOKS FROM HPBOOKS AUTOMOTIVE

HANDBOOKS
Auto Electrical Handbook: 0-89586-238-7
Auto Upholstery & Interiors: 1-55788-265-7
Brake Handbook: 0-89586-232-8
Car Builder's Handbook: 1-55788-278-9
Street Rodder's Handbook: 0-89586-369-3
Turbo Hydra-matic 350 Handbook: 0-89586-051-1
Welder's Handbook: 1-55788-264-9

BODYWORK & PAINTING
Automotive Detailing: 1-55788-288-6
Automotive Paint Handbook: 1-55788-291-6
Fiberglass & Composite Materials: 1-55788-239-8
Metal Fabricator's Handbook: 0-89586-870-9
Paint & Body Handbook: 1-55788-082-4
Sheet Metal Handbook: 0-89586-757-5

INDUCTION
Holley 4150: 0-89586-047-3
Holley Carburetors, Manifolds & Fuel Injection: 1-55788-052-2
Rochester Carburetors: 0-89586-301-4
Turbochargers: 0-89586-135-6
Weber Carburetors: 0-89586-377-4

PERFORMANCE
Aerodynamics For Racing & Performance Cars: 1-55788-267-3
Baja Bugs & Buggies: 0-89586-186-0
Big-Block Chevy Performance: 1-55788-216-9
Big Block Mopar Performance: 1-55788-302-5
Bracket Racing: 1-55788-266-5
Brake Systems: 1-55788-281-9
Camaro Performance: 1-55788-057-3
Chassis Engineering: 1-55788-055-7
Chevrolet Power: 1-55788-087-5
Ford Windsor Small-Block Performance: 1-55788-323-8
Honda/Acura Performance: 1-55788-324-6
High Performance Hardware: 1-55788-304-1
How to Build Tri-Five Chevy Trucks ('55-'57): 1-55788-285-1
How to Hot Rod Big-Block Chevys:0-912656-04-2
How to Hot Rod Small-Block Chevys:0-912656-06-9
How to Hot Rod Small-Block Mopar Engines: 0-89586-479-7
How to Hot Rod VW Engines:0-912656-03-4
How to Make Your Car Handle:0-912656-46-8
John Lingenfelter: Modifying Small-Block Chevy: 1-55788-238-X
Mustang 5.0 Projects: 1-55788-275-4
Mustang Performance ('79–'93): 1-55788-193-6
Mustang Performance 2 ('79–'93): 1-55788-202-9
1001 High Performance Tech Tips: 1-55788-199-5
Performance Ignition Systems: 1-55788-306-8
Performance Wheels & Tires: 1-55788-286-X
Race Car Engineering & Mechanics: 1-55788-064-6
Small-Block Chevy Performance: 1-55788-253-3

ENGINE REBUILDING
Engine Builder's Handbook: 1-55788-245-2
Rebuild Air-Cooled VW Engines: 0-89586-225-5
Rebuild Big-Block Chevy Engines: 0-89586-175-5
Rebuild Big-Block Ford Engines: 0-89586-070-8
Rebuild Big-Block Mopar Engines: 1-55788-190-1
Rebuild Ford V-8 Engines: 0-89586-036-8
Rebuild Small-Block Chevy Engines: 1-55788-029-8
Rebuild Small-Block Ford Engines:0-912656-89-1
Rebuild Small-Block Mopar Engines: 0-89586-128-3

RESTORATION, MAINTENANCE, REPAIR
Camaro Owner's Handbook ('67–'81): 1-55788-301-7
Camaro Restoration Handbook ('67–'81): 0-89586-375-8
Classic Car Restorer's Handbook: 1-55788-194-4
Corvette Weekend Projects ('68–'82): 1-55788-218-5
Mustang Restoration Handbook('64 1/2–'70): 0-89586-402-9
Mustang Weekend Projects ('64–'67): 1-55788-230-4
Mustang Weekend Projects 2 ('68–'70): 1-55788-256-8
Tri-Five Chevy Owner's ('55–'57): 1-55788-285-1

GENERAL REFERENCE
Auto Math:1-55788-020-4
Fabulous Funny Cars: 1-55788-069-7
Guide to GM Muscle Cars: 1-55788-003-4
Stock Cars!: 1-55788-308-4

MARINE
Big-Block Chevy Marine Performance: 1-55788-297-5

HPBOOKS ARE AVAILABLE AT BOOK AND SPECIALTY RETAILERS OR TO ORDER CALL: 1-800-788-6262, ext. 1

HPBooks
A division of Penguin Putnam Inc.
375 Hudson Street
New York, NY 10014